Fundamentals in Organic Geochemistry

Organic Geochemistry is a modern scientific subject characterized by a high transdisciplinarity and located at the edge of chemistry, environmental sciences, geology and biology. Therefore, there is a need for a flexible offer of appropriate academic teaching material (BSc and MSc level) addressed to the variety of students coming originally from different study disciplines. For such a flexible usage the textbook series 'Fundamentals in Organic Geochemistry' consists of different volumes with clear defined aspects and with manageable length. Students as well as lecturers will be able to choose those organic-geochemical topics that are relevant for their individual studies and programs. Hereby, it is the intention to introduce (i) clearly structured and comprehensible knowledge, (ii) process orientated learning and (iii) the complexity of natural geochemical systems. This textbook series covers different aspects of Organic Geochemistry comprising e.g. digenetic pathways from biomolecules to molecular fossils, the chemical characterization of fossil matter, organic geochemistry in environmental sciences, and applied analytical aspects.

Jan Schwarzbauer • Branimir Jovančićević

Isotopes in Organic Geochemistry

 Springer

Jan Schwarzbauer
Institute of Organic Biogeochemistry
in Geo-Systems
RWTH Aachen University
Aachen, Germany

Branimir Jovančićević
Faculty of Chemistry
University of Belgrade
Belgrade, Serbia

ISSN 2199-8647   ISSN 2199-8655   (electronic)
Fundamentals in Organic Geochemistry
ISBN 978-3-031-69306-9   ISBN 978-3-031-69304-5   (eBook)
https://doi.org/10.1007/978-3-031-69304-5

This Springer imprint is published by the registered company Springer Nature Switzerland AG
The registered company address is: Gewerbestrasse 11, 6330 Cham, Switzerland

If disposing of this product, please recycle the paper.

Contents

Chapter 1
Introductory Aspects

Outlook
This chapter introduces the general aspects of isotopes, their composition and variations. Also, the main preconditions for using isotopic analysis in Organic Geochemistry are highlighted.

The most important scientific approach in Organic Geochemistry is identifying, quantifying and interpreting individual organic substances with indicative or marker properties. These molecular aspects are complemented by bulk parameters such as sulfur content, total organic carbon TOC, chemical oxygen demand COD or Rock-Eval parameters such as S1 or S2 values. However, another important aspect is to examine the size dimension below the molecules: the composition of individual atomic nuclei (see Fig. 1.1). Generally, the constitution of atoms differentiates the atomic nuclei and the outer sphere containing the electrons, the electron shells. The atomic nuclei themselves are composed of two types of nucleons: the protons P (positively charged subatomic particles) and neutrons N (neutral subatomic particles). They are subdivided into a more complex structure comprising the quarks (up- and down-quarks) and gluons.

The composition of atomic nuclei has different implications, resulting in different nuclides. Most importantly, the number of protons determines the element, or in other words, atoms with the same number of protons represent the same element independent of the number of neutrons. If the number of protons remains constant but the number of neutrons varies (and consequently also the mass of the nuclides), the resulting atoms represent the group of isotopes (see Fig. 1.2), chemically equal but with different nuclei masses. A corresponding group of nuclides with a constant number of neutrons but varying numbers of protons are called isotones.

© The Author(s), under exclusive license to Springer Nature Switzerland AG 2024
J. Schwarzbauer, B. Jovančićević, *Isotopes in Organic Geochemistry*, Fundamentals in Organic Geochemistry, https://doi.org/10.1007/978-3-031-69304-5_1

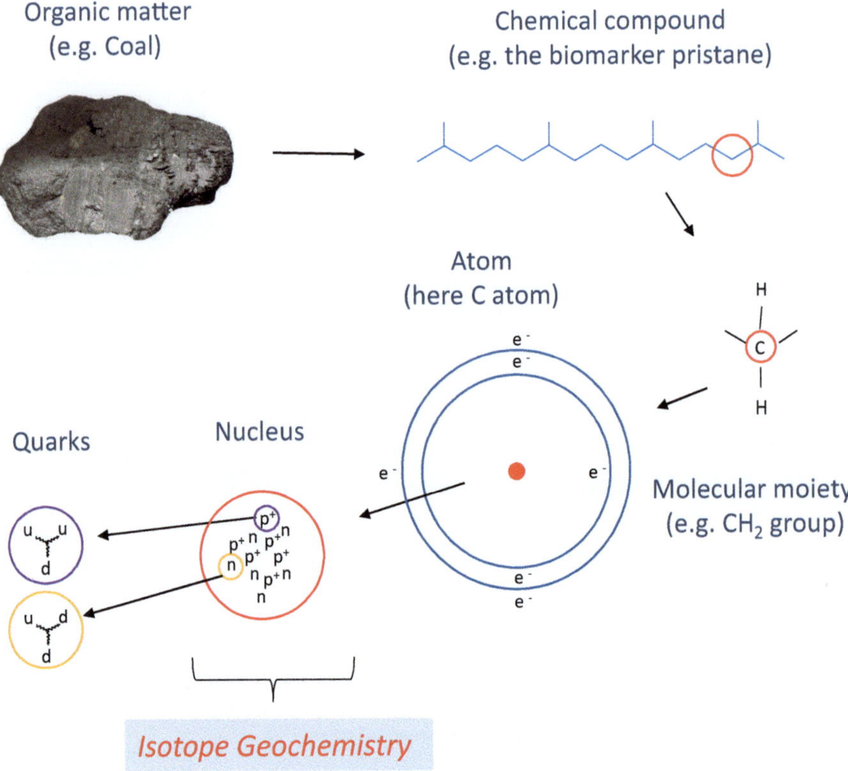

Fig. 1.1 Different size dimensions of organic matter, focusing on atomic nuclei as …

Nuclides with the same mass, meaning the same number of protons and neutrons, form the group of isobars. Additionally, one last group of nuclides, the isodiaphers, exhibit the same excess of neutrons.

Noteworthy, not all possible nuclides exist, and those existing can be divided into the stable ones (stable nuclides) and the unstable ones, the radioactive nuclides. In total, approx. 1900 nuclides exist, of which 267 are stable. Looking at their composition, some stability rules can be deduced. Firstly, comparing the numbers of neutrons and protons, the by far highest quantity of stable nuclides (158 nuclides) consists of even-numbered protons and even-numbered neutrons, followed by the combinations even-numbered P/odd-numbered N (53 nuclides) and odd-numbered P/even-numbered N (50 nuclides). The last combination, all nucleons odd-numbered, is realized only in six nuclides. Secondly, there seems to exist some 'magic numbers' (2, 8, 20, 28, 50, 82) for protons or neutrons, resulting in a high stability or an elevated number of stable nuclides. Thirdly, the Oddo-Harkins rule states that an element with an even number of protons is more abundant than the two adjacent elements with larger or smaller odd proton numbers. This rule describes more or less the above-mentioned preference for stable isotopes with even-numbered protons. Finally, a more general rule can be deduced from the chart of nuclides

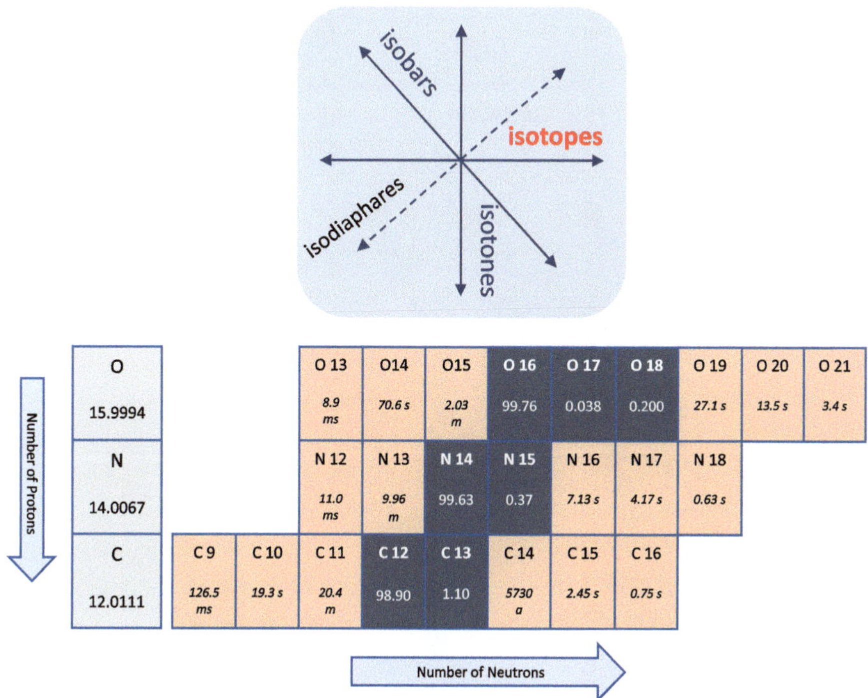

Fig. 1.2 A small section of the *chart of* nuclides comprising the atomic nuclei of carbon, nitrogen and oxygen. Blue-marked are stable, whereas orange-marked nuclides are unstable, meaning radioactive. For stable isotopes, the average relative abundance of the isotope is given (in %), whereas for radioactive nuclides, their half-life time is given (in units from milliseconds to years)

indicating stability for atoms with a balanced number of protons and neutrons up to an atomic number of 20, whereas for higher atomic numbers, the neutrons exceed the protons number. This rule produces the so-called 'line of β-stability' in the nuclides chart.

Looking at the isotopes, the relevant nuclide group in Organic Isotope Geochemistry, solely those appearing in organic molecules are relevant, mainly carbon and hydrogen but also oxygen, nitrogen, sulfur, halogens and phosphor. Carbon isotope analyses represent the most crucial isotopic tool in Organic Geochemistry. A group of stable isotopes for these elements, summarized in Fig. 1.3, are relevant.

The application of isotopic analysis in Organic Geochemistry involves considering several requirements. Firstly, only elements with at least two stable isotopes are relevant. Hence, phosphorus, for example, is not included. Secondly, the abundances of the measured nuclei need to be high enough for a sensitive trace analysis. It is composed of the relative abundance of an isotope as well as the overall abundance of the element itself. As one example, the relative abundance of ^2H is very low (see Fig. 1.3), but hydrogen is one of the most prominent elements in organic compounds. Therefore, hydrogen isotopes are valuable to be measured. Looking at

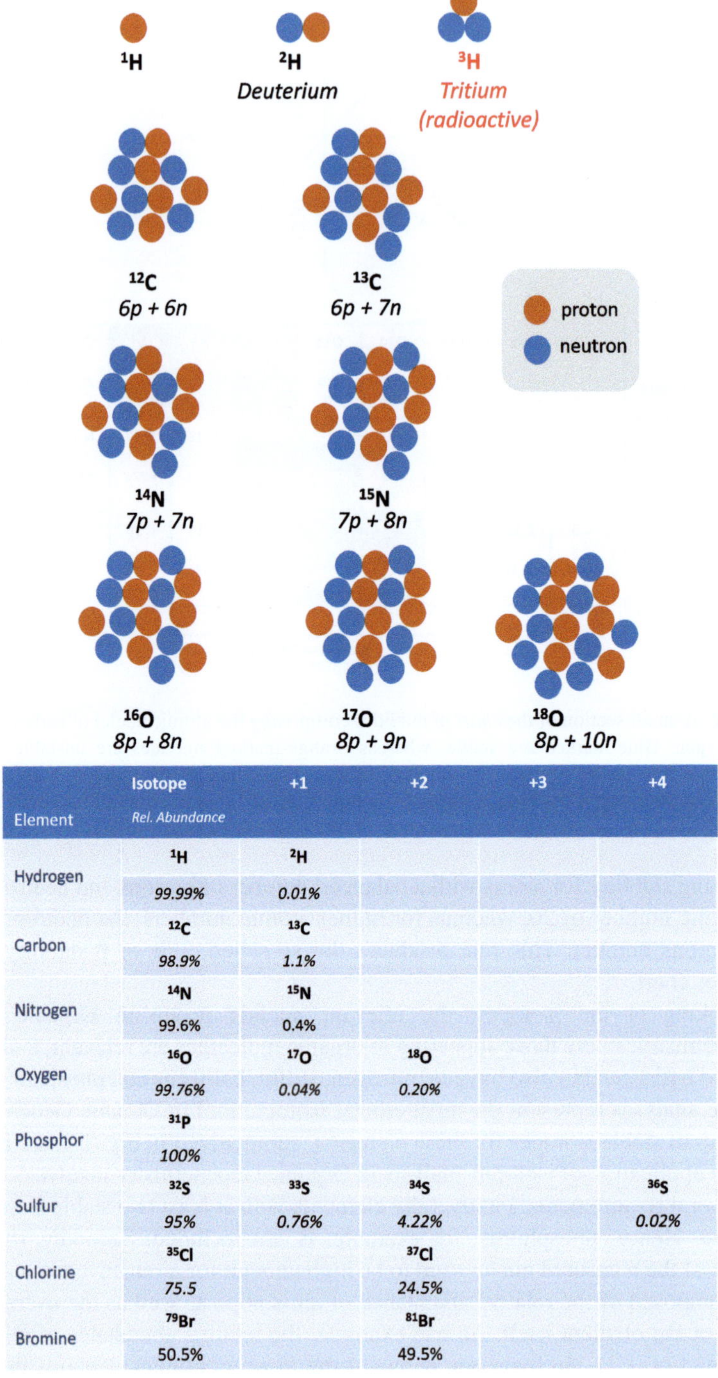

Element	Isotope / Rel. Abundance	+1	+2	+3	+4
Hydrogen	^{1}H / 99.99%	^{2}H / 0.01%			
Carbon	^{12}C / 98.9%	^{13}C / 1.1%			
Nitrogen	^{14}N / 99.6%	^{15}N / 0.4%			
Oxygen	^{16}O / 99.76%	^{17}O / 0.04%	^{18}O / 0.20%		
Phosphor	^{31}P / 100%				
Sulfur	^{32}S / 95%	^{33}S / 0.76%	^{34}S / 4.22%		^{36}S / 0.02%
Chlorine	^{35}Cl / 75.5		^{37}Cl / 24.5%		
Bromine	^{79}Br / 50.5%		^{81}Br / 49.5%		

Fig. 1.3 Stable isotopes relevant in Organic Geochemistry: composition and relative abundances

oxygen isotopes, it is pretty clear that ^{16}O and ^{18}O can be studied, but ^{17}O is analytically neglectable. Finally, although chlorine and bromine are rare elements in natural compounds, their elevated appearance in anthropogenic pollutants (such as dioxins, DDT, PCBs or polybrominated flame retardants), together with the high abundance of ^{37}Cl and ^{81}Br of up to 50%, make them suitable for isotopic analytes in Environmental Organic Geochemistry (see also Schwarzbauer and Jovančićević 2018). However, stable carbon isotope analysis is, by far, the most dominant and essential isotope analysis in Organic Geochemistry.

General Note
The overall abundance of isotopes, composed of their relative abundance and the overall abundance of the element in organic compounds, is the most important precondition for isotope analyses in Organic Geochemistry.

Reference

Schwarzbauer J, Jovančićević B (eds) (2018) Fundamentals in organic geochemistry, vol 3: Organic pollutants in the geosphere. Springer, Cham. 186 pp, ISBN 978-3-319-68937-1

Chapter 2
Isotope Fractionation Processes

Outlook
This chapter refers to principal processes that change the isotopic compositions of individual compounds.

In nature, variations in the isotopic compositions of molecules are evident. These variations are not random but the result of specific processes affecting systematically the isotopic composition. All isotope fractionations are based on processes in which linkages between atoms (chemical or physical bonds) are cleaved or formed.

The basis for the fractionation are different bond strengths or bond energies that depend inter alia on the atomic weight of the involved atoms. This undoubtly accounts for different elements and isotopes. However, if the bond energies vary for bonds in which different isotopes are involved, the corresponding reaction kinetics also vary based on the resulting different activation energies (see Fig. 2.1). The differences in energies between the initial and transition states, which exhibit higher energy in particular due to bond cleavage, represent the activation energy of the reaction. The activation energy directly influences the reaction's kinetic or velocity. This follows the well-known Arrhenius equation (see Fig. 2.1). If the activation energy for a reaction with a heavier isotope is higher due to the lower energy of the ground state, the resulting reaction kinetic is slower.

In summary, the bond energies with heavier isotopes are higher, leading to slower reaction kinetics. Hence, an enrichment of the process products with lighter isotopes can be observed during processes in which a corresponding reaction is participated. This is an isotopic fractionation due to a so-called kinetic isotope effect.

In principle, primary and secondary kinetic isotope effects can be distinguished. A primary fractionation is based on a bond cleavage or formation in which the relevant isotopes are directly involved. A secondary fractionation can be observed if bonds are cleaved or formed at positions near the isotopically different atoms.

© The Author(s), under exclusive license to Springer Nature
Switzerland AG 2024
J. Schwarzbauer, B. Jovančićević, *Isotopes in Organic Geochemistry*,
Fundamentals in Organic Geochemistry,
https://doi.org/10.1007/978-3-031-69304-5_2

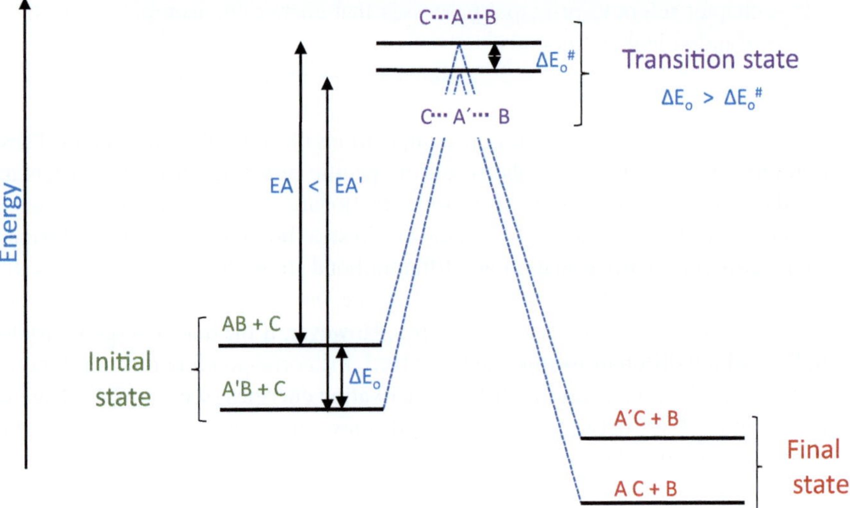

Fig. 2.1 General overview of reaction kinetics comprising types of reaction (first, second … order), the energy profiles for reactions with different isotopes (A and A′), as well as the relation between activation energy and reaction kinetics as given by Arrhenius

The isotopic shift of secondary effects is usually smaller than primary isotope effects and can sometimes lead (but rarely) to an inverse kinetic effect.

Aside from the kinetic isotope fractionation, there is also an equilibrium isotope fractionation, existing dominantly for the partition between two phases (e.g., gaseous and liquid). Also, the extent of isotopic shifts is smaller than that of the primary kinetic isotope effects. As a further difference, the equilibrium effects are reversible, whereas kinetic fractionations are mostly irreversible. However, more importantly, in Organic Geochemistry, only the kinetic isotope effect is relevant, meaning that the equilibrium effect will be neglected in this textbook.

The extent of the fractionation is based inter alia on the relative differences of the isotope masses. Since the relative mass difference for stable isotopes is highest for $^1H/^2H$ by doubling the mass of the lighter isotope, the kinetic isotope effect is also the highest. As shown in Table 2.1, the ratio of the rate constants k for the relation $^1H/^2H$ or H/D is 18, whereas corresponding values for carbon, nitrogen or oxygen isotopes are around 1.2.

These kinetic isotope effects affect the physical properties of molecules and their reactions. Since the isotopic effect is highest for the element hydrogen, the implications for physicochemical properties depending on intermolecular interactions are most prominent. This refers, for example, to the differences in boiling point, density or viscosity for H_2O and D_2O, but also thermodynamic parameters differ according to the different bond energies and intermolecular forces (see Table 2.2). In the case of molecules consisting of two or more elements with stable isotopes, the physico-chemical differences vary for the isotope composition of both elements. This is also exemplified in Table 2.1 for considering both H and D, as well as ^{16}O and ^{18}O in water molecules.

To illustrate the kinetic isotope effect on chemical transformations, the ratio of ^{12}C and ^{14}C species in typical organic reactions are summarized in Table 2.3.

Table 2.1 Calculated maximum kinetic isotope effects at 25 °C

Isotopes	k_1/k_2
H/D	18
$^{12}C/^{13}C$	1.25
$^{14}N/^{15}N$	1.14
$^{16}O/^{18}O$	1.19
$^{127}I/^{131}I$	1.02

Table 2.2 Influence of isotope composition on physical and thermodynamic properties of water

	$H_2^{16}O$	$D_2^{16}O$	$H_2^{18}O$
Boiling point (°C)	100	101.42	100.14
Melting point (°C)	0.00	3.81	0.28
Density (g/cm³, at 20 °C)	0.997	1.1051	1.1106
Viscosity (centipoise at 25 °C)	1.002	1.247	1.056
Free enthalpy (kJ/mol, gaseous)	−228.65	−233.30	
Heat of evaporation (J/mol at 25 °C)	44.011	45.388	

Data partly adapted from Hoefs (2018)

Table 2.3 The kinetic isotope effect in two different organic reactions with respect to ^{12}C and ^{14}C

Reaction	k_{12}/k_{14}
Decarboxylation of acetic acid	1.06
Hydrolysis of methyl benzoate	1.16

Although the radioactive and heavier isotope ^{14}C (compared to ^{13}C) has been chosen for demonstration, the resulting effect is low but detectable.

> **General Rules for Isotope Effects**
> - The binding energy with a heavier isotope is higher
> - The reaction rate of the heavier isotope is slightly lower
> - In chemical equilibria, the heavier isotope occurs preferentially in the chemical form that has the higher heat of formation

The ratio of stable isotopes in organic compounds also varies to some extent due to isotope fractionation processes. To accurately follow such alterations, a quantifying parameter is needed. Principally, changes in isotopic composition are given as the ratio of two different stable isotopes. The $^{13}C/^{12}C$ ratio is the most important parameter, followed by the $^{2}H/^{1}H$ or D/H ratio.

Since these variations are minimal, the usage of absolute values is not convenient, so the alterations are given in relation to a standard reference material. Each element has its standard reference material, as summarized in Fig. 2.1. For carbon isotopes, the global $^{13}C/^{12}C$ ratio has an average value of 0.011235, the $^{13}C/^{12}C$ ratio of the corresponding standard reference material (a mineral, the Pee Dee Belemnite, standardized as so-called Vienna Pee Dee Belemnite or VPDB) is 0.011180 ($\pm$16). These two values indicate that the isotope ratio difference is usually very small. Therefore, the isotopic values, expressed in the so-called δ-notation, are given as per mille shifts in relation to the standard (see Fig. 2.2). For stable carbon, the δ-notation is applied as $\delta^{13}C$ (VPDB). Note at a more negative value represents an enrichment of the lighter isotope.

> **Example of $\delta^{13}C$ (VPDB) Notation**
> If an absolute isotopic ratio $^{13}C/^{12}C$ of 1.1015% is measured, the corresponding δ-value can be calculated according to the equation in Fig. 2.2. by taking the difference of the measured value with the standard value (VPDB) and dividing it by the standard value. The resulting value needs to be multiplied by 1000 for the ‰ notation:
>
> | 1. Step: Difference | $0.011015 - 0.011180 = -0.000165$ |
> |---|---|
> | 2. Step: Division | $-0.000165/0.011180 = -0.01476$ |
> | 3. Step: Multiplication | $-0.01476 \times 1000 = -14.76$ |
> | | **Result: $\delta^{13}C$ (VPDB) = −14.76‰** |

$$\delta X(‰) = \left(\frac{R_{Sample} - R_{Standard}}{R_{Standard}} \right) * 1000$$

$$\text{with} \quad R = \frac{heavier\ isotope}{lighter\ isotope}$$

Element	Isotope ratio (E_h/E_l)	Average proportion (%)	Accepted value ($\times 10^6$) [°C]	Standard reference material
C	$^{13}C/^{12}C$	1.110 : 98.890	11180	Vienna Pee Dee Belemnite (VPDB), carbonate
N	$^{15}N/^{14}N$	0.3663 : 99.63	3676.5	N2, atmospheric
O	$^{18}O/^{16}O$ $^{17}O/^{16}O$	0.2040 : 99.76 0.0383 : 99.76	2005.20 379.9	Vienna Standard Mean Ocean Water (VSMOW)
		alternative	2067.2 385.9	Vienna Pee Dee Belemnite (VPDB), carbonate
H	$^{2}H/^{1}H$ or D/H	0.0156 : 99.98	155.76	Vienna Standard Mean Ocean Water (VSMOW)
S	$^{34}S/^{32}S$	4.2200 : 95.02	44162.6	Vienna Canyon Diablo Troilit (VCDT)

Fig. 2.2 Principal definition of the δ notation and isotope ratios of the corresponding standard reference material for the most common elements in Organic Geochemistry (data adapted from Hoefs 2018; Hayes 1983; De Wit et al. 1980; Ding et al. 2001; Meija et al. 2016)

2.1   Qualitative Basics

As already mentioned, isotope ratio shifts result from physical or chemical processes. Two fundamental and geoscientifically relevant processes are related to both types, respectively, and affect different elements. They are illustrated in Fig. 2.3 and are discussed in detail in the following.

The global water cycle consists of two fundamental transfer processes: evaporation and precipitation. Although they are physical processes, the core of the phase transfer is the linkage and formation of intermolecular bonds, dominantly the hydrogen bonds. Correspondingly, the processes influence the isotopic composition due to slightly different kinetics for lighter and heavier isotopes. Since both oxygen and hydrogen are involved in the hydrogen bonds, also both elements are isotopically affected. For example, oxygen evaporation shifts the isotopic composition roughly by approx. 10‰ (VSMOW) (see Fig. 2.3).

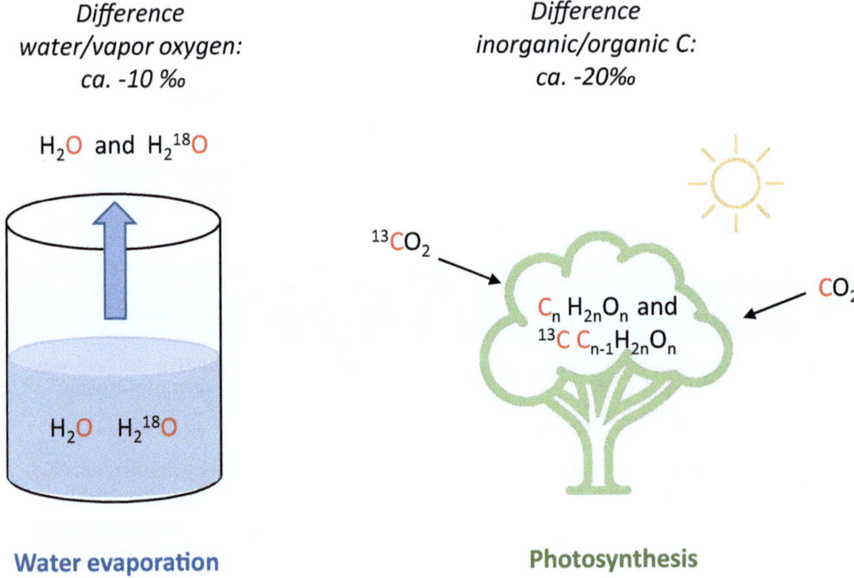

Fig. 2.3 Two natural processes causing significant isotopic shifts

The same accounts for the hydrogen isotopes. As already figured out here, the isotopic shift is more pronounced and scientifically more intensively described with an enrichment of ^1H during evaporation. Following the water cycle along the main pathway from the marine systems, where especially in warm to tropical regions, partly high amounts of water become vaporized, towards the terrestrial environment, characterized by enhanced rainfall, not only a one-step cycle can be observed. A multistep cascade of evaporation and precipitation leads to a systematic enrichment of lighter isotopes (see Fig. 2.4).

This effect is visible even on a continental scale. The more inland in continental systems, the lighter the water resources, both vapor and rain (including corresponding surface water systems). This is exemplified for North America in Fig. 2.5, where the δD values shift from $-20‰$ in the Gulf of Mexico to $-140‰$ in Northern Canada. The same accounts also for the oxygen isotopes enriched by the lighter ^{16}O isotope in rainwater more inland.

The second phenomenon of a 'global' isotope fractionation is photosynthesis, the basic natural reaction for producing organic matter on Earth. This reaction chain generally starts with inorganic CO_2 and ends with the organic compound glucose (see Fig. 2.6).

Here, the fundamental isotopic shift is linked with incorporating carbon dioxide and CO_2 fixation (more detailed information, see Chapter 1.2 in Schwarzbauer and Jovančićević 2016). This chemical reaction differs for various plant types,

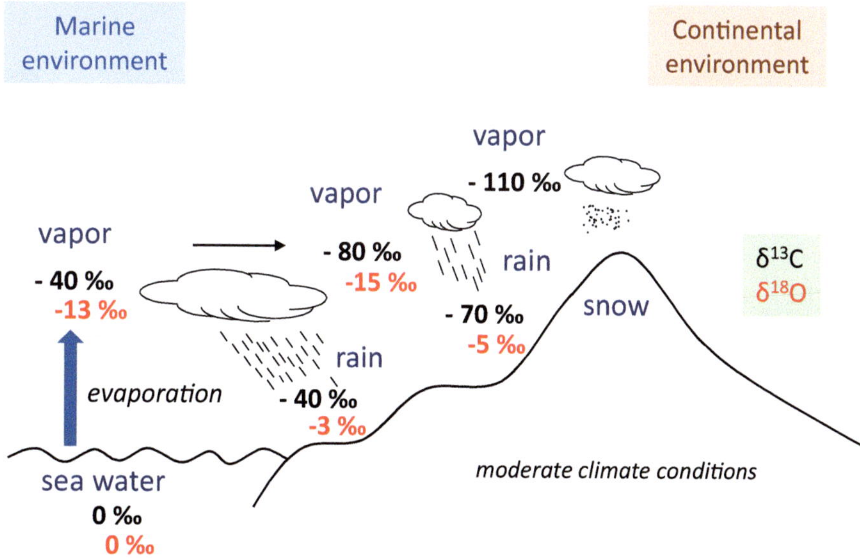

Fig. 2.4 Principal shift of δD values in water (rain, vapor) resulting from evaporation followed by cascading rainfall from marine to continental regions

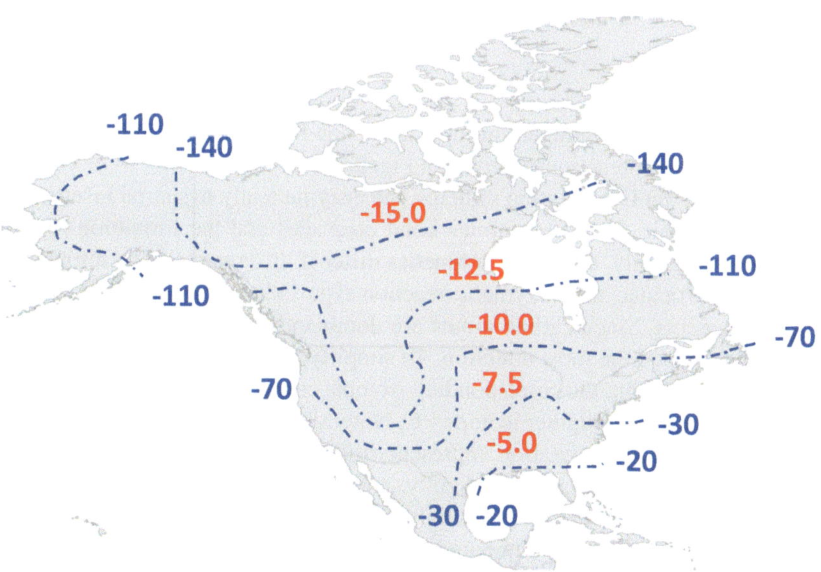

Fig. 2.5 Schematic distribution of δD (blue) and δ^{18}O (red) values (‰) in rainwater of North America (according to White 2005, chapter 9 and Dansgaard 1964)

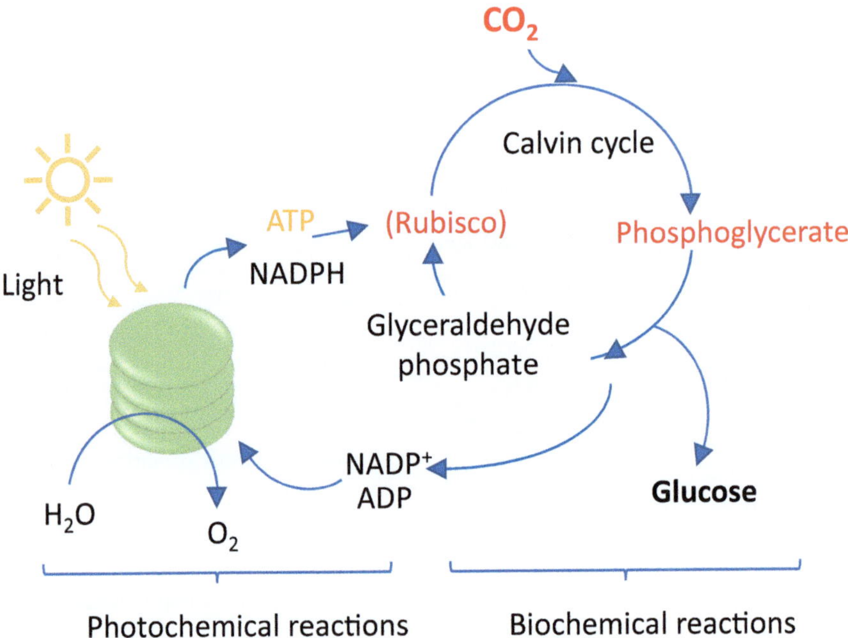

Fig. 2.6 A simplified scheme of photosynthesis in C3 plants. The isotopically most important reaction step, the CO$_2$ fixation in the Calvin cycle, is highlighted in red

particularly for the two dominant groups, the C3 and C4 plants. A simplified summary of both fixation reactions is given in Fig. 2.7.

In both dominant CO$_2$-fixation systems, the enzymatically triggered insertion of CO$_2$ into an organic molecule implies bond cleavages and the formation of new linkages. Consequently, the reaction kinetics differ if different carbon isotopes are involved. In particular, the enzymatic reaction type induces an enhanced isotope fractionation effect. Since carbon atoms are dominantly involved, the carbon isotope shift has gained the most attention. To simplify, during the change from inorganic to organic carbon, the corresponding isotopic composition shifts significantly to a higher proportion of lighter isotope ^{12}C. Note, since both mechanisms, the PEP and RuBisCo-related CO$_2$ fixation, represent different chemical reaction pathways, the resulting isotopic shift differs slightly, and therefore, the carbon pool in C3 plants is principally characterized by a lighter carbon isotope composition (meaning more negative δ^{13}C values) as compared to the C4 plants.

The resulting quantitative implications from a more global perspective are discussed in the following chapter.

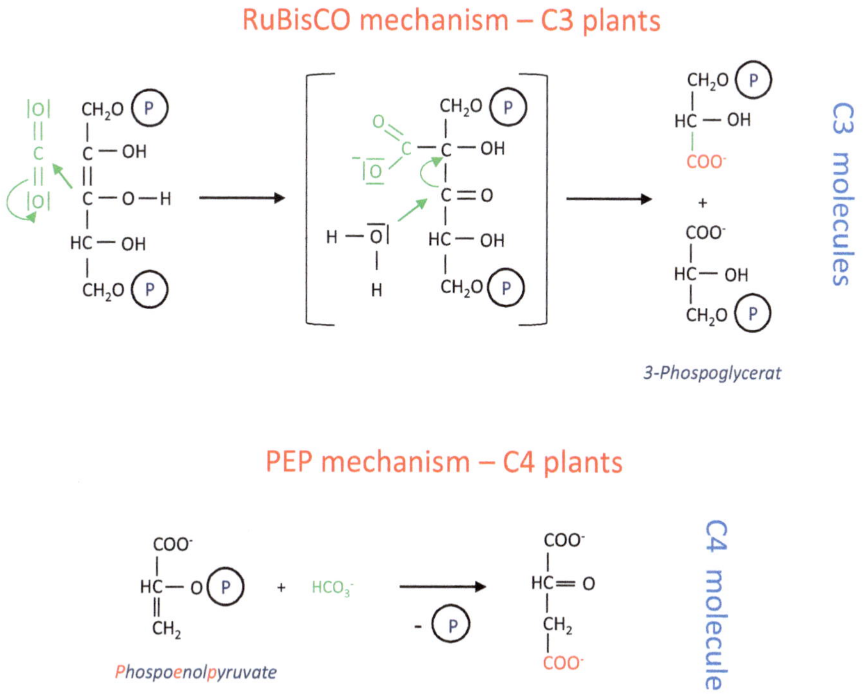

RuBisCO mechanism – C3 plants

C3 molecules

3-Phospoglycerat

PEP mechanism – C4 plants

C4 molecule

Phospoenolpyruvate

Fig. 2.7 The key reactions of CO_2 fixation in C3 and C4 plants using RuBisCo and PEP systems, respectively

2.2   Quantitative Aspects

2.2.1   The Rayleigh Equation

Since isotope shifts are based on physicochemical principles, the extent of shifting can be calculated. Such quantitative calculations of isotope fractionation processes are commonly based on the Rayleigh equation. This equation was initially developed for the fractional distillation of two mixed liquids, but the principal approach has also been introduced to isotope shift calculations.

If we apply this approach to a chemical reaction, educt and product (e.g., in a biotic degradation process) need to be distinguished. Two different but complementary terms can express the differences in isotope composition or its fractionation: the fractionation factor α or the enrichment factor ε. Detailed definitions are given in Fig. 2.8 (Eqs. I and II). The enrichment factor is the relation between the relative isotope compositions **R** (s. Fig. 2.2) of product to educt. The enrichment factor also

Isotope fractionation between two compounds expressed as:

$$\alpha_{p-r} = \frac{R_{product}}{R_{reactant}} = \frac{10^{-3}\delta_p + 1}{10^{-3}\delta_r + 1} \quad (I) \qquad\qquad \varepsilon_{p-r} = \left(\frac{R_{product}}{R_{reactant}} - 1\right) \cdot 1000 \quad (II)$$

$$= (\alpha - 1) \cdot 1000 \; [\text{‰}] \quad (III)$$

fractionation factor α   enrichment factor ε

$$\frac{R_t}{R_0} = \left[f\frac{(1+R_0)}{(1+R_t)}\right]^{(\alpha-1)} = \left[f\frac{(1+R_0)}{(1+R_t)}\right]^{\varepsilon} \quad (IV)$$

$$f = \frac{L_t + H_t}{L_0 + H_0} = \frac{L_t(1+R_t)}{L_0(1+R_0)} \quad (V)$$

Relationship between substrate concentration
change and isotope fractionation

If H+L ≈ L or 1+R_f ≈ 1+R_o
Simplified to:

$$\frac{R_t}{R_0} = f^{(\alpha-1)} \quad (VI)$$

$$\ln\left(\frac{R_t}{R_0}\right) = (\alpha - 1)\ln f = \frac{\varepsilon}{1000}\ln f \quad (VII)$$

If |ε|<20‰
Simplified to:

$$\delta_{r,t} - \delta_{r,0} = \varepsilon \ln f \quad (VIII)$$

R – ratio heavy isotope to light isotope (at t=0 and t)
f – remaining fraction of the reactant at time t
L_0 – concentrations light isotope
H_0 – concentrations heavy isotope

Fig. 2.8 Principal approaches and basic terms for quantitation of isotope shifts (after Schmidt et al. 2004)

compares this relation with 100%. Since both parameters describe the same process but from slightly different points of view, they have an interrelationship, as given in Fig. 2.8 (Eqs. IV and V).

However, the isotope shift develops over reaction time coupled with the change in concentrations (amounts) of the reaction species, product and educt. In isotope shift calculation, the so-called remaining fraction of the educt f is usually used to

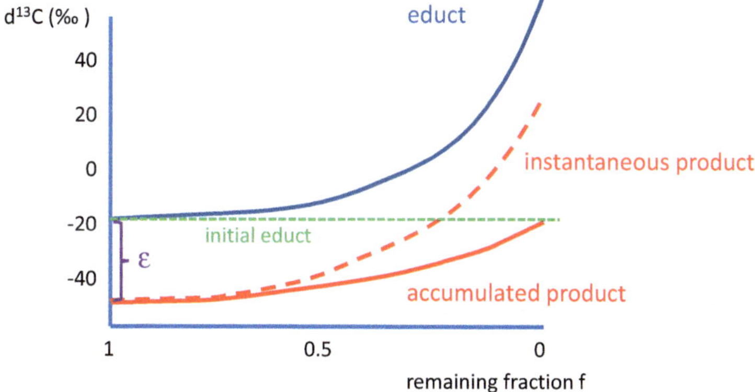

Fig. 2.9 Simplified development of isotope ratios for educts and products during a reaction (simplified after Schmidt et al. 2004)

express the mass changes over time. The basic equation for quantitation of isotope shifts is also given in Fig. 2.8 (Eq. III). However, isotope compositions in natural abundances are characterized by low values for heavier isotopes. Further on, the isotope shifts themselves are relatively low.

As a consequence of both aspects, Eq. (III) can be simplified by the Rayleigh approach to the direct relation of isotope ratios **R** at different time points, the remaining fraction f and the fractionation factor α or the enrichment factor ε (Eqs. VI and VII). Further simplification can be applied in the case of low enrichment factors, which are usually observable in natural processes. The most simplified relation of isotope composition and reaction results is given in Eq. (VIII).

The changes in isotope compositions during natural reactions are certainly specific regarding reaction species, environment, catalysis etc. The principal development of isotope ratios during a reaction is schematically exemplified in Fig. 2.9.

Principally, the corresponding basic parameters α and ε are unique and need to be determined individually for each reaction. The simplified Eq. (VIII) can be used to experimentally determine these parameters. However, it needs to be mentioned that such a determination is only an approximation. On the one hand, the exact same conditions as for the observed processes in nature need to be considered, and the above-mentioned simplifications certainly provide uncertainties to some extent. Finally, the calculations are valid strictly only for reactions in closed systems with the formation of only one product and no subsequent reaction step. Despite such limitations, the described principal quantitative approach has been applied successfully in numerous studies using dynamic isotope shifts for following natural processes in biosynthesis and biodegradation. Noteworthy, the calculation of isotope shifts is not restricted to chemical reactions but is also valid for physical processes such as vaporization.

2.2.2 *Global Aspects*

As shown, individual reactions and physico-chemical interaction can induce isotope fractionation effects that can be quantified. Although these isotopic shifts are defined well for individual processes of molecules, the isotopic fractionation is more complex in natural systems. The natural conditions are more comprehensive even for characterized molecular processes, e.g., photosynthesis. The CO_2 fixation is incorporated in consecutive steps, leading to a superimposition of different effects on the isotope composition. These steps differ for the type of CO_2 fixation, particularly by comparing C3 and C4 plants. In C3 plants, the transfer of CO_2 to the enzyme system, as well as the fixation, are located in one cell. Here, the photosynthetic isotope fractionation consists of the enzymatic fractionation at the RuBisCo system ($\delta^{13}C$ approx. $-27\%o$) and the CO_2 diffusion towards the cell (approx. $-4\%o$). These chemical and physical processes do not contribute additively, meaning the resulting isotope effect needs to be calculated in a more complex way. For C4 plants, it becomes more difficult. There is not only one but two diffusion steps since the overall CO_2 fixation happens in two separate cells. The resulting isotope shift is composed of the CO_2 fixation at the PEP system, the diffusion towards the mesophyll cell and, as a second transport pathway, the diffusion to the bundle sheath cells. Also, here, the resulting isotope fractionation is not only the addition of the individual shifts.

To make the situation more complex, the concentration and modification of the available CO_2 also affect the resulting isotope fractionation. For instance, in aquatic plants, the diffusion of CO_2 partially as HCO_3^- is a limiting step. Moreover, the isotopic composition of the CO_2 itself also determines the absolute resulting carbon isotope ratio after fixation. A (schematic) calculation for C3 and C4 plants is illustrated in Fig. 2.10 to exemplify the increasing complexity.

Following this, ecosystem conditions also influence isotope fractionation from a molecular level towards more complex natural systems. This comprises inter alia the differences between marine and terrestrial systems, the nutrient supply and further parameters. Hence, it is evident that a higher variation of isotope shifts is visible in nature. Therefore, concerning Organic Geochemistry, not only are individual isotopic shifts on a molecular level worth following, but more general consequences

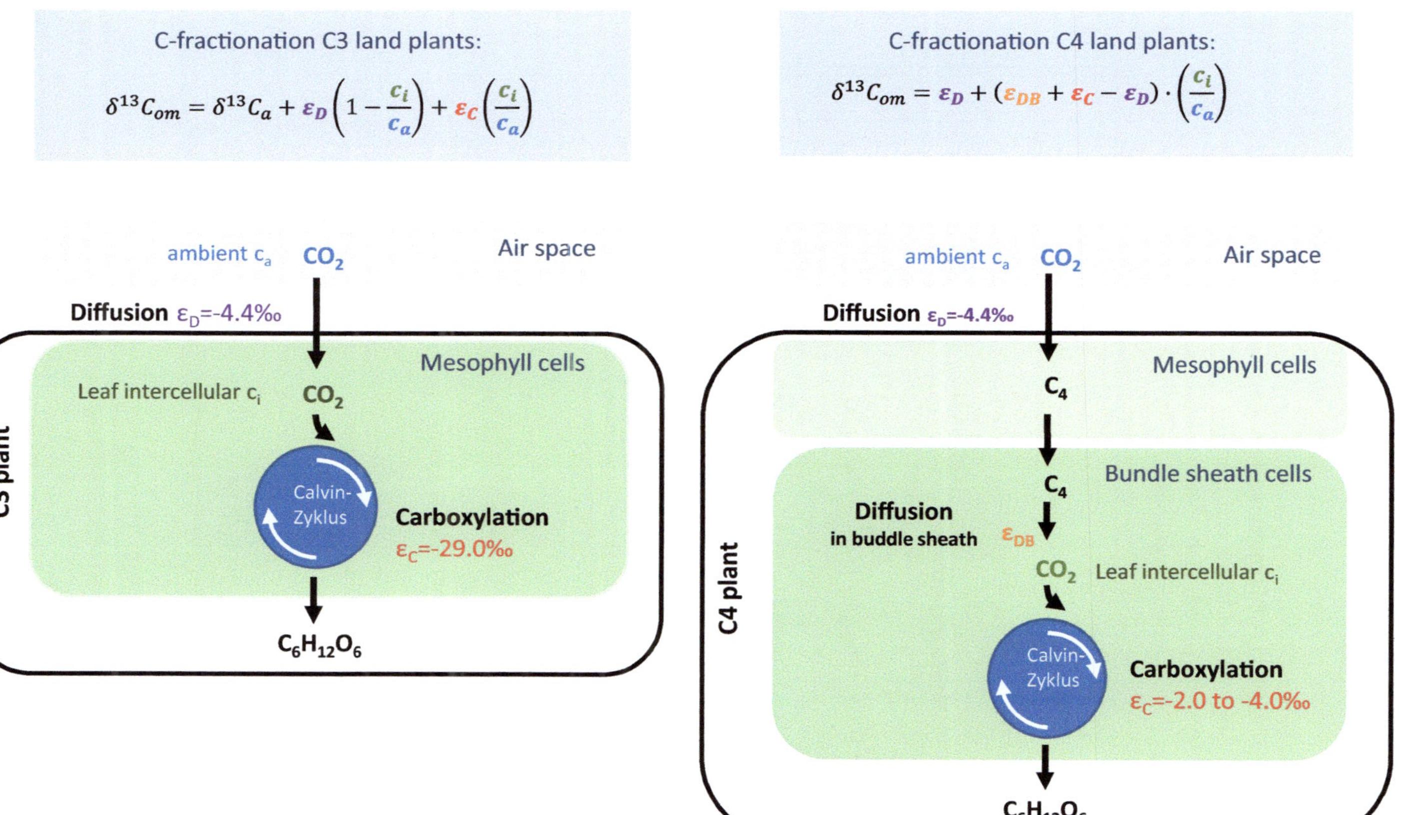

Fig. 2.10 Visualization of the different chemical and physical processes affecting the isotope shift during carbon fixation in C3 and C4 plants (modified after White 2005; Hayes 1983; Farquhar et al. 1989; Goericke et al. 1994)

and principal trends observable on a macroscopic, even global dimension, are important and interesting.

Example of Natural $\delta^{13}N$ Fractionation: The Aquatic Food Chain
An isotope shift related to the nitrogen isotopes can be observed along the aquatic ecosystems' food chain. The systematic enrichment of heavier ^{15}N results from ^{14}N depletion during assimilation and excretion.

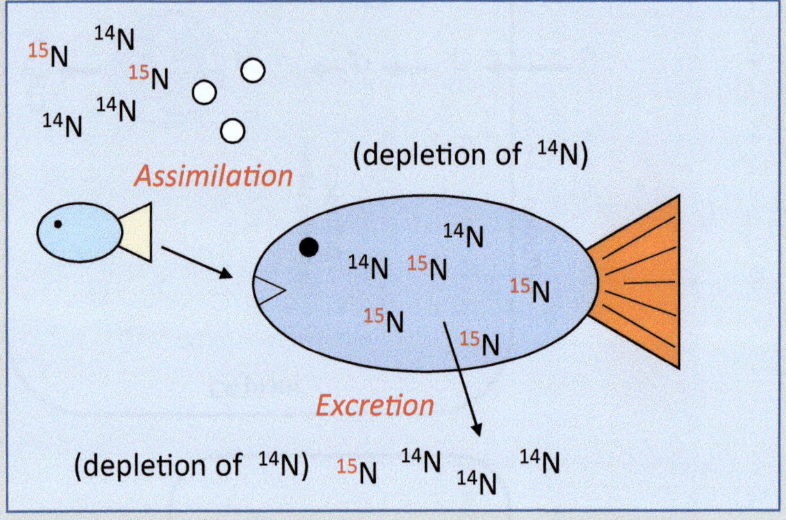

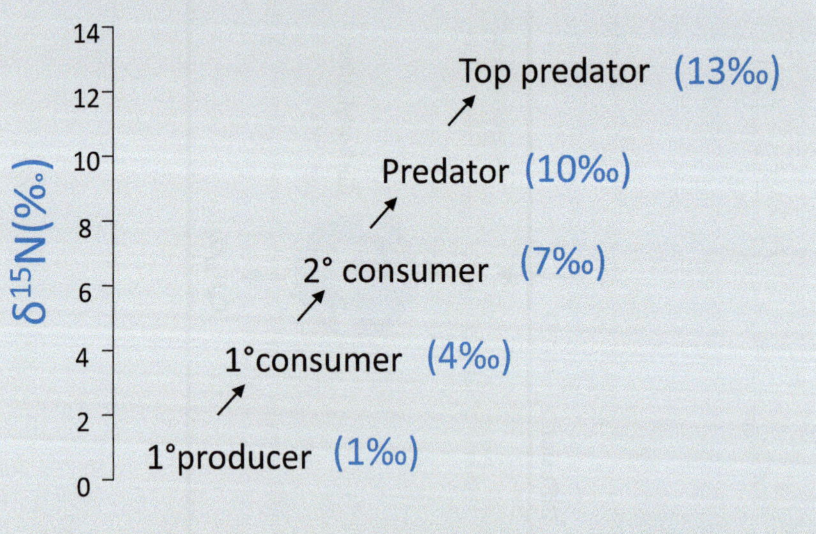

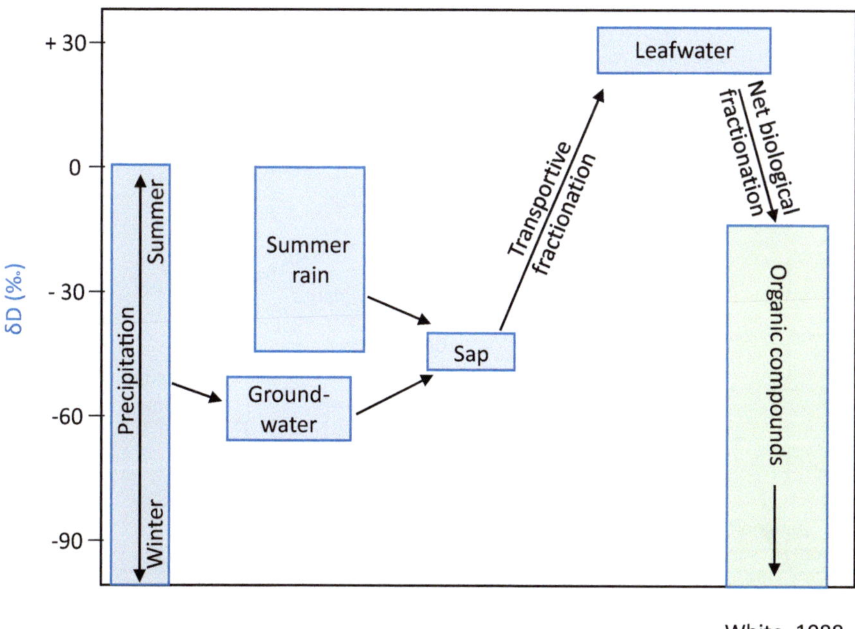

Fig. 2.11 Schematic variety of δD values in different natural matter (adapted from and modified after White 2005)

For the hydrogen isotopes, a principal global distribution is exemplified by some pools of water resources and organic material in Fig. 2.11. Here, the range of δD values covers approx. 120‰, which is more as compared to $\delta^{13}C$ values in natural systems. In particular, the high variety in organic matter is a characteristic of hydrogen isotopes. This is related to the high reactivity of hydrogen in various functional groups or molecular moieties in principle. Hence, a rapid exchange of hydrogen atoms is often observable, leading to an intensive isotope shifting and, consequently, a wider range of δD values.

However, for Organic Geochemistry, the global fractionation effects of carbon are more fundamental (see Fig. 2.12). Here, especially, the before-mentioned fractionation during photosynthesis allows us to distinguish pools of organisms more precisely. The most prominent differentiation can be observed for the already mentioned C3 and C4 plants, as well as further plant species such as CAM plants. As stated, the inorganic (e.g., marine carbonates, atmospheric CO_2) and organic carbon fractions are also clearly separated. As an indicator of how specific isotope shifts can be, the human breath of European and American people differs remarkably in terms of carbon isotope composition. Concerning fossil matter, the $\delta^{13}C$ values cover a wider range as it is subject to numerous chemical and physical alterations over time. Nevertheless, the principal level clearly points to the biogenic origin of fossil matter.

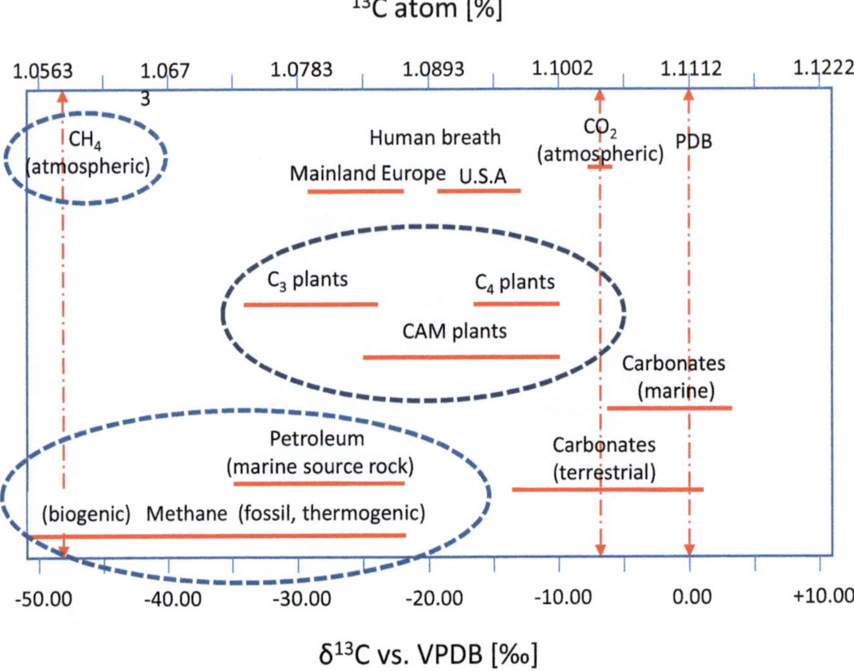

Fig. 2.12 Schematic variety of δ^{13}C values in different natural matter (according to Brand 1996)

Transferred to a more geoscientific perspective, the isotope balance of carbon in the geosphere can be attributed not only to individual pools but also to exchange processes between these carbon accumulations. This is illustrated schematically in Fig. 2.13. Here, a more or less dynamic transfer, exchange and conversion of carbon is related to reservoirs in the bio- and geosphere. As an important aspect, the isotope variations always need to be interpreted in relation to the overall amount of carbon accumulated in the reservoirs or systems.

Finally, Fig. 2.14 roughly estimates and summarizes an overview of the geoscientific relevant reservoirs. Here, the overall amount and the isotope composition are reflected. The important influence of photosynthesis on the global distribution of carbon isotopes is once more evident. From an organic geochemical point of view, organic and inorganic carbon are clearly distinguishable not only by their molecular structure but also by their isotope composition. Hence, both molecular and isotopic characterization are essential tools in Organic Geochemistry.

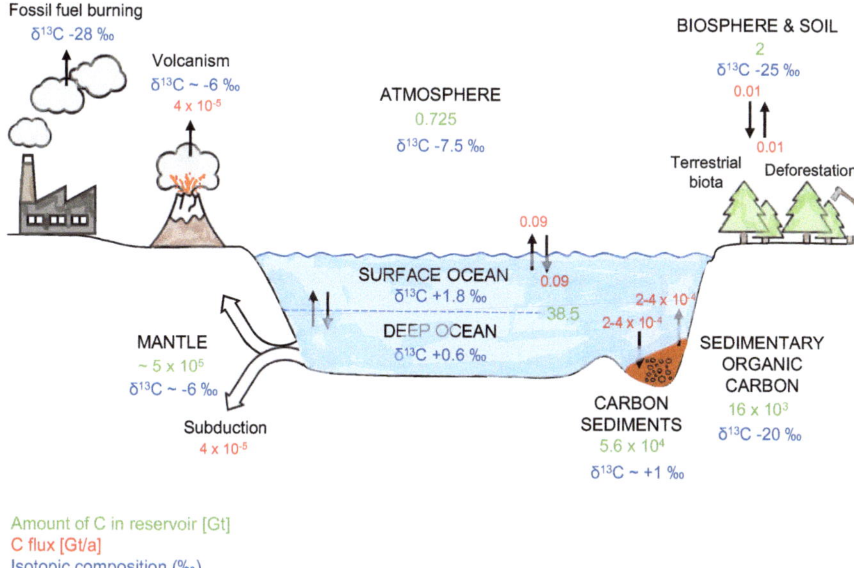

Fig. 2.13 More detailed carbon cycle including long- and short-term elements and corresponding carbon isotope signatures

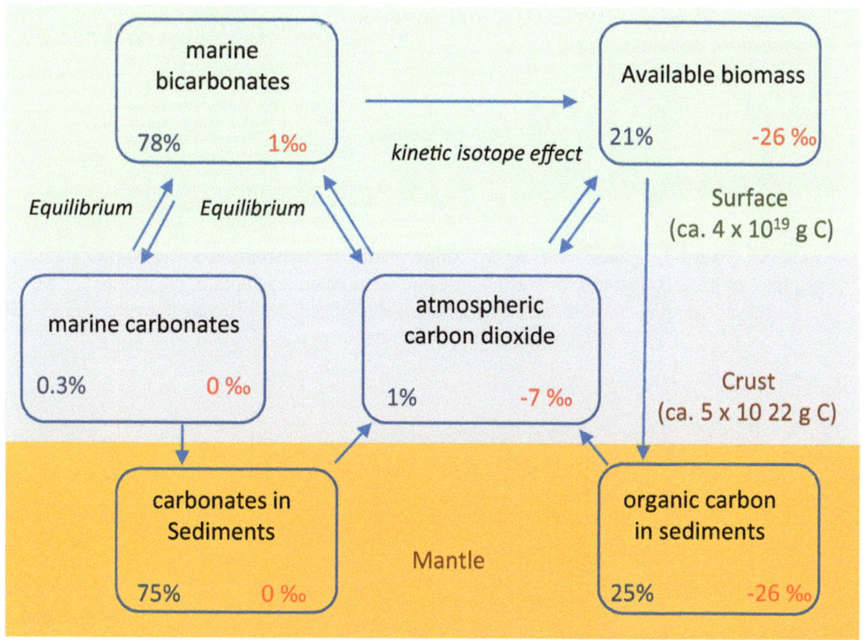

Fig. 2.14 The distribution of carbon in the geosphere with respect to the relative mass balance in the mantle, as well as crust/surface (in %, blue), and the average isotope composition (in ‰ δ13C, red)

References

Brand WA (1996) High precision isotope ratio monitoring techniques in mass spectrometry. J Mass Spectrom 31:225–235

Dansgaard W (1964) Stable isotopes in precipitation. Tellus 16:436–468

De Wit JC, Straaten CM, Mook WG (1980) Determination of the absolute hydrogen isotope ratio of V-SMOW and SLAP. Geostand Geoanal Res 4:33–36

Ding T, Valkiers S, Kipphardt H, De Brievre P, Taylor PDP, Gonfiantini R, Krouse R (2001) Calibrated sulfur isotope abundances ratios of three IAEA sulfur isotope reference materials and V-CDT with a reassessment of the atomic weight of sulfur. Geochim Cosmochim Acta 65:2433–2437

Farquhar GD, Ehleringer JR, Hubick KT (1989) Carbon isotope discrimination and photosynthesis. Annual Rev Plant Physiol Plant Mol Biol 40:503–537

Goericke R, Montoya JP, Fry B (1994) Physiology of isotopic fractionation in algae and cyanobacteria. In: Lajtha K, Michener RH (eds) Stable Isotopes in Ecology and Environmental Science. Wiley, pp 187–221

Hayes JM (1983) Practice and principles of isotopic measurements in organic geochemistry. In: Organic geochemistry of contemporaneous and ancient sediments, vol 5. Great Lake Section SEPM, Bloomington, IN, pp 1–5.31

Hoefs J (2018) Stable isotope geochemistry, 8th edn. Springer, Berlin. 208 p, ISBN 978-3-319-78526-4

Meija J, Coplen TB, Berglund M et al (2016) Atomic weights of the elements 2013 (IUPAC technical report). Pure Appl Chem 88:265–291

Schmidt TC, Zwank L, Elsner M, Berg M, Meckenstock RU, Haderlein SB (2004) Compound-specific stable carbon isotope analysis of organic contaminants in natural environments: a critical review of the state of art, prospects, and future challenges. Anal Bioanal Chem 378:283–300

Schwarzbauer J, Jovančićević B (eds) (2016) Fundamentals in organic geochemistry, Vol 2: From biomolecules to chemofossils. Springer, Cham. 160 pp, ISBN 978-3-319-27241-2

White WM (2005) Geochemistry. Wiley, Oxford

Further Reading

Schmidt TC, Zwank L, Elsner M, Berg M, Meckenstock RU, Haderlein SB (2004) Compound-specific stable carbon isotope analysis of organic contaminants in natural environments: a critical review of the state of art, prospects, and future challenges. Anal Bioanal Chem 378:283–300

Chapter 3
Analytical Methods

Outlook
This chapter deals with the analytical basics for stable isotope determination. Here, the analysis of bulk material and compound-specific measurements need to be distinguished. Both approaches are mainly based on mass spectrometry, either in a static mode or continuous flow approach.

3.1   Bulk Stable Isotope Analysis

The analysis of stable isotope compositions is based on mass spectrometry since the isotopes differ in atomic weight. Indeed, mass spectrometry has been developed primarily for isotope analyses. The first application of mass spectrometry was the separation of ^{20}Ne and ^{22}Ne by J.J. Thomsen in 1913, representing a fundamental step in the isotope theory.

However, organic molecules consist of a specific structure with a carbon skeleton and substituents that disables the direct measurement of stable isotopes. The clue for isotope analysis is the conversion of organic molecules or bulk organic matter into CO_2 and further easily detectable gases for a simplified mass spectrometric detection of the stable isotopes. Certainly, there are some important preconditions. As the most important one, the conversion must be complete, mentioned, e.g., for stable carbon isotope analysis, that all organic carbon atoms need to be converted to CO_2. That implies that the formation of byproducts needs to be prevented; in the case of carbon isotopes, e.g., this applies to the formation of CO. The step for conversion into measurable gases is oxidation or combustion. As mentioned, carbon is measured by CO_2, but for nitrogen, the conversion forms the target gas N_2 and

© The Author(s), under exclusive license to Springer Nature
Switzerland AG 2024
J. Schwarzbauer, B. Jovančićević, *Isotopes in Organic Geochemistry*,
Fundamentals in Organic Geochemistry,
https://doi.org/10.1007/978-3-031-69304-5_3

nitrogen oxides; hence, a reduction process is added. Hydrogen isotopes are measured by H_2.

Looking deeper into stable carbon isotope analyses, it needs to be mentioned that the measurement of $^{12}CO_2$ vs. $^{13}CO_2$ by the m/z values of 44 and 45 is insufficient. The stable isotopes of oxygen interfere, as illustrated in Fig. 3.1, leading to the superimposition, particularly for the m/z 45 signal. This effect by the isotopomers can also be corrected by measuring the m/z 46 signal to calculate the $^{12}C^{16}O^{17}O$ fraction.

The technical realization of bulk stable carbon analysis is trickier, as described. In principle, it converts bulk material, e.g., petroleum, organic-rich sediments, and natural gas, into CO_2 and its mass spectrometric detection. This approach was introduced to Organic Geochemistry in the middle of the twentieth century. It was expanded by pretreatment steps to separate the bulk material into sub-fractions. Examples include the measurement of fractionated extracts or purified gas samples. Separation was conducted by liquid chromatography (e.g., the separation of oils

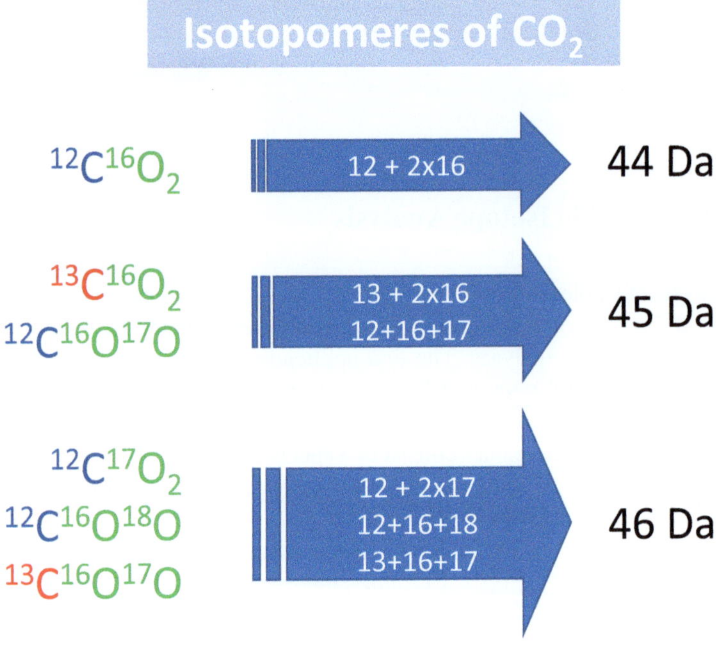

Fig. 3.1 The main isotopomers of CO_2, their masses and the corresponding co-detection in mass spectrometry

into aliphatic and aromatic hydrocarbons, as well as NSO fractions) or on mol sieves in case of gases (separating methane from higher alkanes).

Excursus: Example of Traditional Separation Approaches for Stable Carbon Analysis

Below is an older, more historical separation approach for natural gases. However, it nicely illustrates the efforts needed to separate different carbon fractions (CO_2, methane, and higher homologs) to assess the individual organic carbon isotope composition.

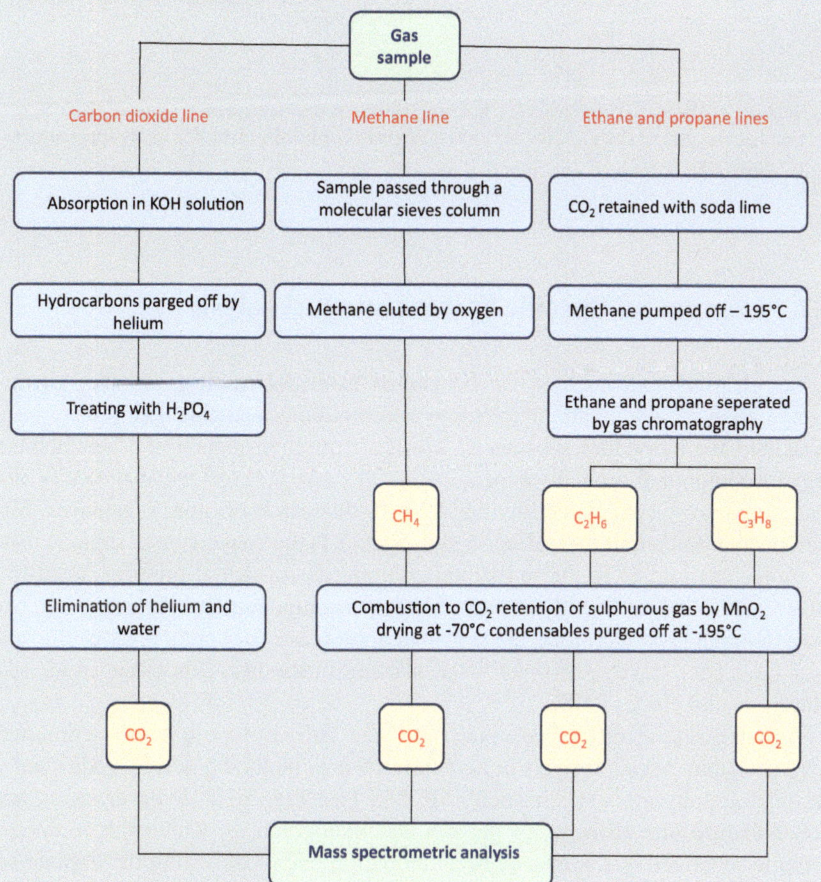

Schematic representation of fraction conversion: (1) methane and (2) ethane and propane into CO_2 for mass spectrometric analysis (according to Colombo et al. 1964)

In the 1980s, a technical separation approach, as shown in the following figure, was used for bulk stable carbon analysis of oil, bitumen, or fractionated extracts.

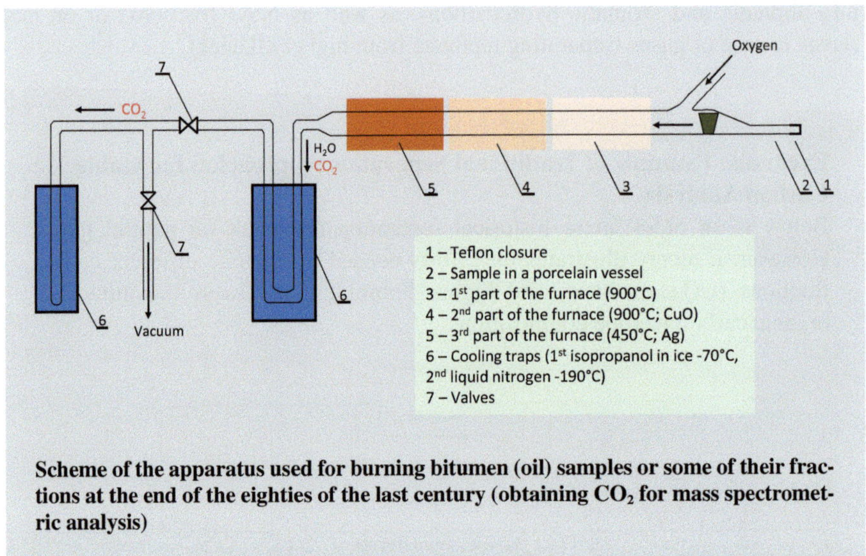

Scheme of the apparatus used for burning bitumen (oil) samples or some of their fractions at the end of the eighties of the last century (obtaining CO_2 for mass spectrometric analysis)

3.2 Compound-Specific Stable Isotope Analysis

Although much information can be gained from determining the stable isotope composition of bulk matter, much more information is available if isotope ratios are measured for individual substances. However, in environment and geochemistry, organic compounds typically appear in complex mixtures and are analytically available in extracts accompanied by hundreds to thousands of other substances. Since isotope analysis requires (as described in Sect. 3.1) the conversion of organic matter or molecules into CO_2, the individual compounds in natural mixtures need to be either isolated or sufficiently separated before a compound-specific analysis. This is a challenging task; therefore, the principal approaches for compound-specific analyses have been developed much later, starting in the later 70s (Sano et al. 1976; Matthews and Hayes 1978).

As an essential difference concerning the bulk isotope analysis, compound-specific stable carbon analysis is a continuous-flow method based on traditional gas chromatography-mass spectrometry GC/MS (see Fig. 3.2). However, as an additional feature, the effluents of the gas chromatograph, in which the compounds become separated, are subjected to an oxidation oven, converting the organic substances to CO_2 before the mass spectrometric stable carbon isotope measurement. The combination detects stable carbon isotope composition of the separated, meaning individual compounds. Consequently, this method is named *compound-specific isotope analyses* (CSIA), and the corresponding mass spectrometric system is an *isotope ratio monitoring* MS (irm MS). However, it should be noted that the quality of information is quite different compared to the results obtained from traditional

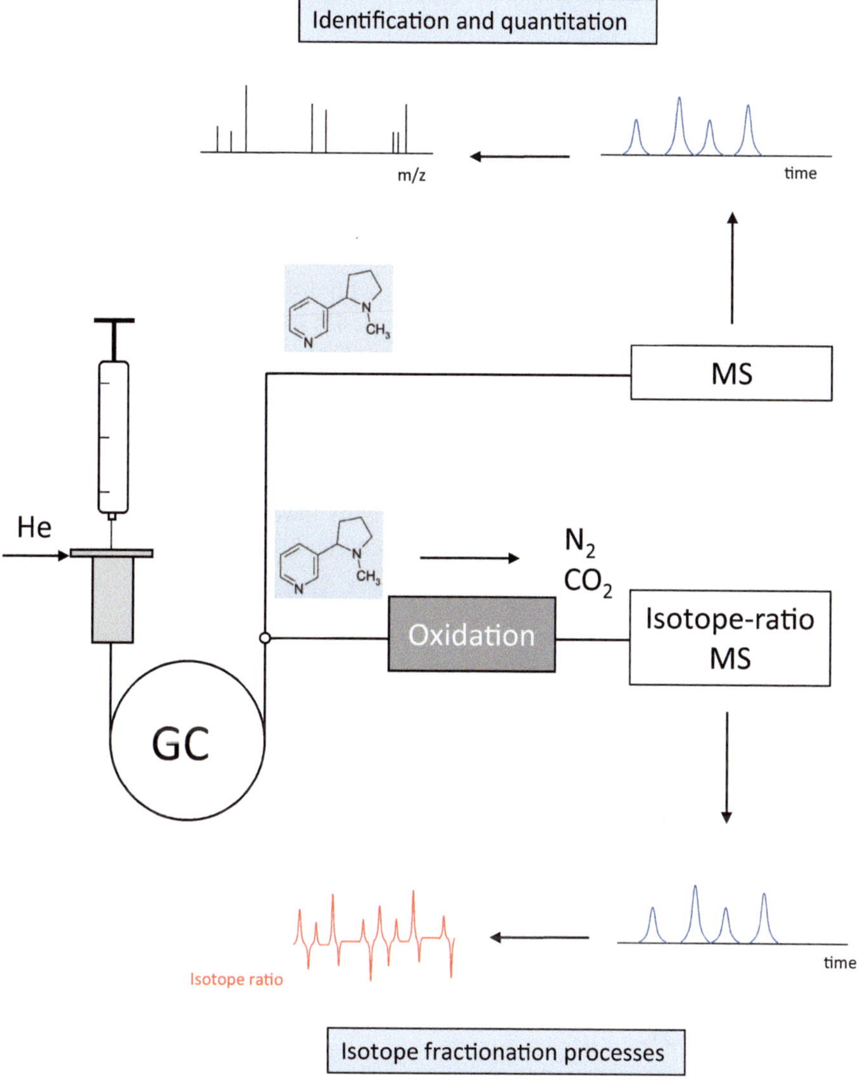

Fig. 3.2 The principal differences between traditional gas chromatography-mass spectrometry and stable isotope analyses

mass spectrometry (see Fig. 3.2). All molecular structure details, measured by molecule fragments as illustrated in the standard mass spectra, are removed in the oxidation oven. For a complete characterization of organic compounds, including identification, quantitation and isotope composition, both standard MS and irmMS are needed.

This technique is not restricted to carbon isotopes. Still, some modifications are needed for nitrogen or hydrogen isotope analyses, as mentioned in Sect. 3.1, since other gases are generated, interfering with the MS detection (Merritt et al. 1995). In the case of nitrogen, the oxidation of organic matter leads not exclusively to the elemental nitrogen (N_2) but also to NO and NO_2. Hence, in the nitrogen mode, the oxidized gases become reduced in a second oven, in which the transformation of nitrogen oxides to elemental nitrogen occurs (see Fig. 3.3).

Independent from the element, irmMS measurements produce ion chromatograms of individual isotopic masses. For stable carbon isotope analysis, these chromatograms represent the abundances of CO_2 with different isotopic compositions (m/z 44, m/z 45 and m/z 46 for correction of m/z 45; see Sect. 3.1), including isotopomers (as summarized in Fig. 3.1). These ion chromatograms are the base for the calculation of isotope ratios. The peak areas represent the overall amount of the respective isotopes, but the special conditions of gas chromatographic elution need to be considered. The retention and, consequently, the retention time depends inter alia on the masses of the substances. In the case of isomers, the compounds exhibiting the heavier isotope composition have a slightly lower retention time or in other words, they elute earlier from a GC column (see Fig. 3.4). Taking this phenomenon

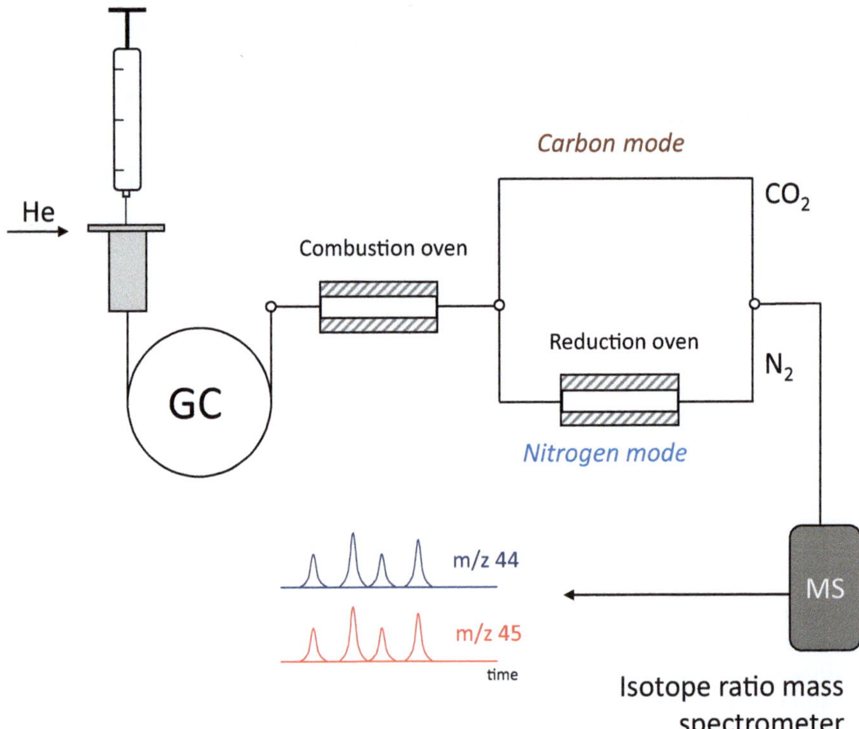

Fig. 3.3 Scheme of an isotope mass spectrometer operating in the compound-specific mode

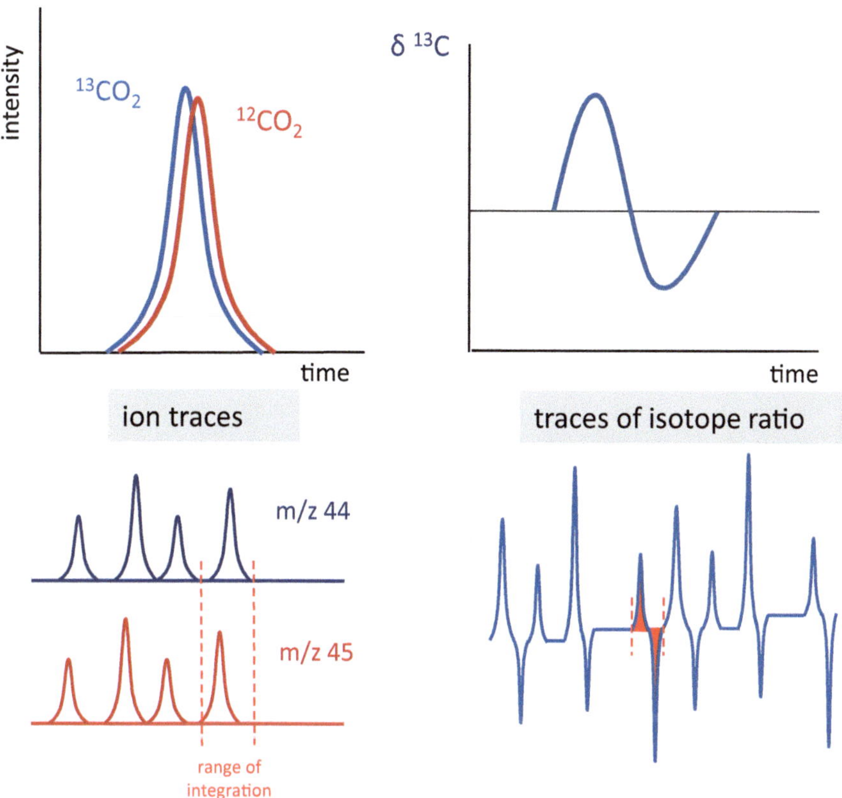

Fig. 3.4 Gas chromatographic behavior of $^{13}C/^{12}C$ ratios forming the 'swing' (upper graphics) and the consequent implications for a full-range integration covering the whole GC peak, as illustrated in the graphics below

into account, the result of calculating the ratio of $^{13}C/^{12}C$ for each data point in GC-irmMS results in a so-called 'swing' as illustrated in Fig. 3.4. Due to an earlier appearance of ^{13}C (or enrichment of ^{13}C in earlier eluting compounds) the relative composition is dominated by ^{13}C resulting in high $^{13}C/^{12}C$ values. After some time, this relation shifts towards higher contributions of ^{12}C, leading to an inverse shift of the $^{13}C/^{12}C$ ratio. These shifts form the 'swing' shape of the $^{13}C/^{12}C$ ratios over the whole GC peak for a measured substance. This 'swing'-behavior has some important implications. Firstly, the accurate isotope fraction within one GC signal can be calculated only by integrating the entire peak from the ion chromatogram (see Fig. 3.4). Secondly and more importantly, a correct calculation of $^{13}C/^{12}C$ ratios is only possible for chromatographically isolated substances, meaning fully separated GC peaks. If there is an overlap or superimposition of another substance, the isotope ratios interfere, leading to incorrect $^{13}C/^{12}C$ ratios. In the case of incompletely GC-separated peaks for two components, the lighter isotope composition of the first eluting substance would influence the heavier fraction of the subsequently eluting compound. In other words, isotope compositions of compounds can be determined by GC-irmMS only in case of completely separated GC signals.

Excursus: Precision in irmMS

As for all quantitative measurements, quantifying isotope ratios exhibits analytical errors. Since the variation of isotope compositions is within a very small quantitative interval, deviations during the measurement represent an important issue in compound-specific isotope analyses. Important parameters influencing the accuracy are inter alia the combustion temperature and conditions in combustion ovens. Further on, the extent of deviation depends on the absolute amount of the substances. The lower the measurement amount, the higher the deviation. Consequently, reliable CSIA analyses are performed at least as duplicates but often as triplicates.

The principal order of precision depends on the isotope and the overall isotope shift. Compound-specific hydrogen isotope measurements with a range of more than a hundred ‰ SMOW exhibit a level of precision around 5‰, whereas, for carbon isotopes ($\delta^{13}C$ values range around 30‰VPDB), a deviation of 0.2‰ is common.

To exemplify the level of deviations, the following figure shows some measured carbon isotope values and corresponding deviations of pollutants along a section of the river Rhine, Germany:

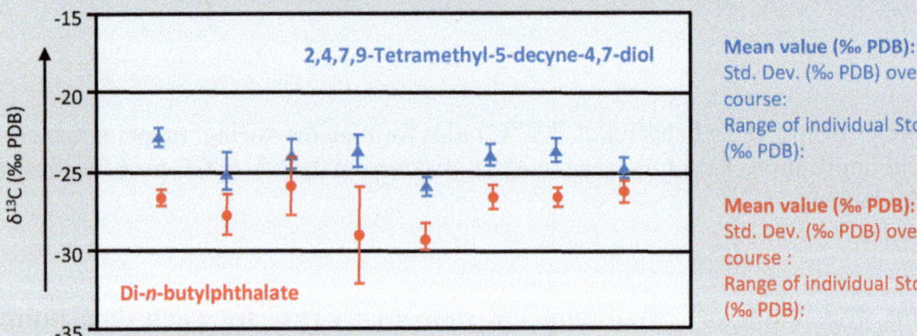

$\delta^{13}C$-values of selected contaminants in water samples from the Rhine River—on the right side, the average value of the isotope ratios with the corresponding standard deviations obtained from the individual sampling locations and along the river course (adapted from Schwarzbauer et al. 2005). Here, it can be observed that the variation along the river is similar to the analytical deviations. Hence, no clear trend can be deduced from this data set

General Note

GC-irmMS is a powerful tool for determining the isotope compositions of individual compounds. However, some limitations need to be considered. On the one hand, information on individual structural properties gets lost, while on the other hand, a high chromatographic resolution is needed for a correct calculation.

References

Colombo U, Gazzarrini F, Confiantini R, Sironi G, Tongiorgi E (1964) Measurements of C13/C12 isotope ratios on Italian natural gases and their geochemical interpretation. In: Hobson GD, Louis MC (eds) Advances in Organic Geochemistry. Pergamon Press, pp 279–292

Matthews DE, Hayes JM (1978) Isotope-ratio-monitoring gas chromatography-mass spectrometry. Anal Chem 50:1465–1473

Merritt DA, Freeman KH, Ricci MP, Studley SA, Hayes JM (1995) Performance and optimization of a combustion interface for isotope ratio monitoring gas chromatography/mass spectrometry. Anal Chem 67:2461–2473

Sano M, Yotsui Y, Abe H, Sasaki S (1976) A new technique for the detection of metabolites labelled by the isotope ^{13}C using mass fragmentography. Biomed Mass Spectrom 3:1–3

Schwarzbauer J, Dsikowitzky L, Heim S, Littke R (2005) Determination of $^{13}C/^{12}C$-ratios of anthropogenic organic contaminants in river water samples by GC-irmMS. Int J Environ Anal Chem 85:349–364

Chapter 4
Aspects of Compound-Specific Isotope Analysis Apart from Geosciences

Outlook

This chapter describes various examples of the application of isotope analyses in analytical disciplines other than Organic Geochemistry, including food, drug, and doping analysis.

Stable isotope analysis is a powerful tool in nearly all fields of Organic Geochemistry, but it has also been successfully applied in other scientific and applied analytical areas. Interestingly, the general approaches of applications are always very similar; hence, it is worth giving a brief overview of some examples from non-geoscientific practices.

The main approach is using isotope data for an unambiguous linkage of products and origin, as well as between two different pools of organic matter. This linkage is based on specific isotope fingerprints of raw materials, e.g., water, feed or production materials. The reasons for isotopic differences are already addressed to some extent in Chap. 2. For example, systematic isotopic shifts in water concerning the continental location have been described. Furthermore, isotope characterization of plant material depends inter alia on botanical type (CO_2 fixation) and ecological conditions. These variances often generate individual isotopic signatures that can be used for fingerprinting. Such a straightforward approach has been applied in different disciplines; some examples are introduced in the following.

Isotope-based fingerprinting is a principal strategy common in food sciences and food protection. Table 4.1 summarizes already established methods. In the food industry, provenance is a very sensitive and critical property that often needs an unambiguous linkage between food products and the geographical origin of ingredients or production sites.

© The Author(s), under exclusive license to Springer Nature Switzerland AG 2024
J. Schwarzbauer, B. Jovančićević, *Isotopes in Organic Geochemistry*,
Fundamentals in Organic Geochemistry,
https://doi.org/10.1007/978-3-031-69304-5_4

Table 4.1 Official EU analytical methods employing isotope data and related projects

Product	Official method/research project	Application
Wine	UE # 2676/90 822/97, 440/2003, OIV, Wine-DB, GLYCEROL	Chaptalization, geographical provenance, watering
Spirits	OIV, BEVABS	Botanical origin, ethanol
Sugar	SUGAR ^{18}O	Botanical origin
Honey	AOAC method 999.41	Addition of sugars
Dairy and cheese	MILK	Geographical provenance
Fruit juice	AOAC method 995.17, SUGAR ^{18}O	Addition of sugars, dilution
Oil and fat	MEDEO	Origin
Fish	COFAWS	Geographical provenance; wild vs. cultured
Vinegar	OIV	Botanical origin

Adapted from Calderone et al. (2003)

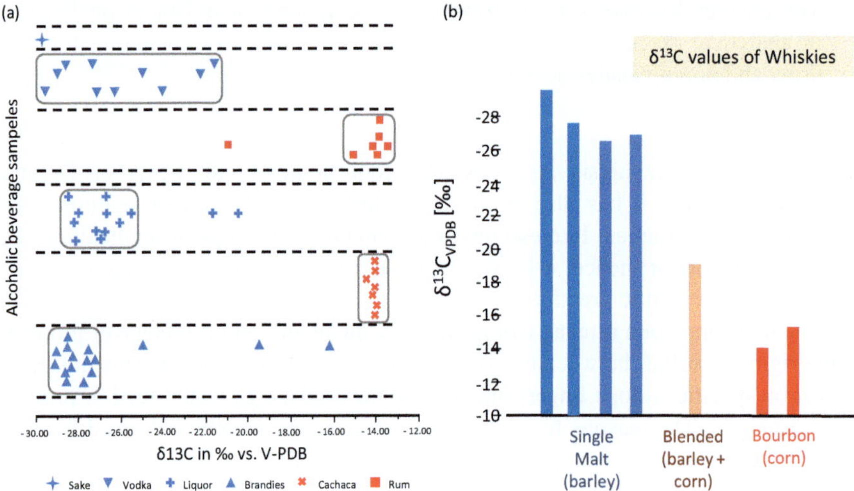

Fig. 4.1 (**a**) Ranges of δ¹³C values for various types of spirits and (**b**) δ¹³C values of different types of whisky (adapted and modified from Jochmann et al. 2009)

The first set of more detailed examples covers the field of spirits and alcoholic beverages. In spirits, the alcohol is derived from different biological sources with varying isotope characteristics. A systematic differentiation of various types of spirits is exemplified in Fig. 4.1. Based on the botanic origin of ethanol, corn- and sugar cane-derived products (e.g., rum or tequila) can be differentiated from spirits produced from rice, wheat or potato (e.g., vodka or sake). Both plant groups exhibit different types of CO_2 fixation, representing either C4 (corn, sugar cane) or C3 plants (rice, wheat, potato). Their different isotope signatures (see Chap. 2) are preserved in the isotope composition of the corresponding ethanol and, consequently, in the spirit product. Noteworthy, the isotopic signature can be superimposed by further ingredients like herbs or additives (e.g., anise spirits), shifting the bulk carbon isotope signal. Here, a compound-specific analysis allows a more accurate classification.

Whisky production is also based on different raw plant materials, either barley in the case of Scottish single malt (C3 plant-derived) or corn in the case of American Bourbon (C4 plant-derived). Consequently, the carbon isotope signature of a whisky sample can be used to verify or falsify a proposed origin, as illustrated in Fig. 4.1b. Therefore, the isotope signature can be used more accurately to distinguish differences on a product level.

The same strategy can be applied to beer. Malt and wheat as main ingredients influence the isotopic composition of beer as measured in dry extracts (see Fig. 4.2). Here, a two-dimensional correlation of carbon and nitrogen isotope composition characterizes individual beer sorts. Notably, the carbon isotope variations discriminate different groups. Moreover, adding polysaccharides from corn and rice significantly shifts towards lower $\delta^{13}C$ values since these additives are derived from C4 plants.

The link between isotopes and individual constituents has also been used for product characterization of caffeine-containing beverages and drinks. Here, biogenic and synthetic caffeine can be differentiated isotopically. For example, caffeine in tea and energy drinks differs isotopically, with $\delta^{13}C$ values from −25 to −32‰ (VPDB) or from −33 to −38‰ (VPDB), respectively. The values point to an isotopically lighter product from technical synthesis than natural caffeine. These differences are used to identify mislabeled products, as exemplified in Fig. 4.3.

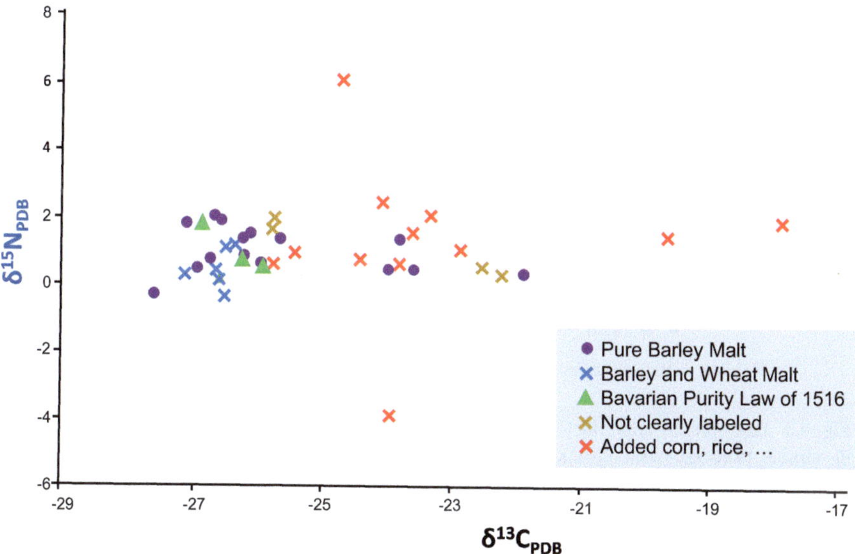

Fig. 4.2 Discrimination of various sorts of beer based on a two-dimensional correlation of carbon and nitrogen fingerprints. Note the shift toward heavier signals for modified beer by adding corn or rice components (adapted and modified from Ehleringer et al. 2007)

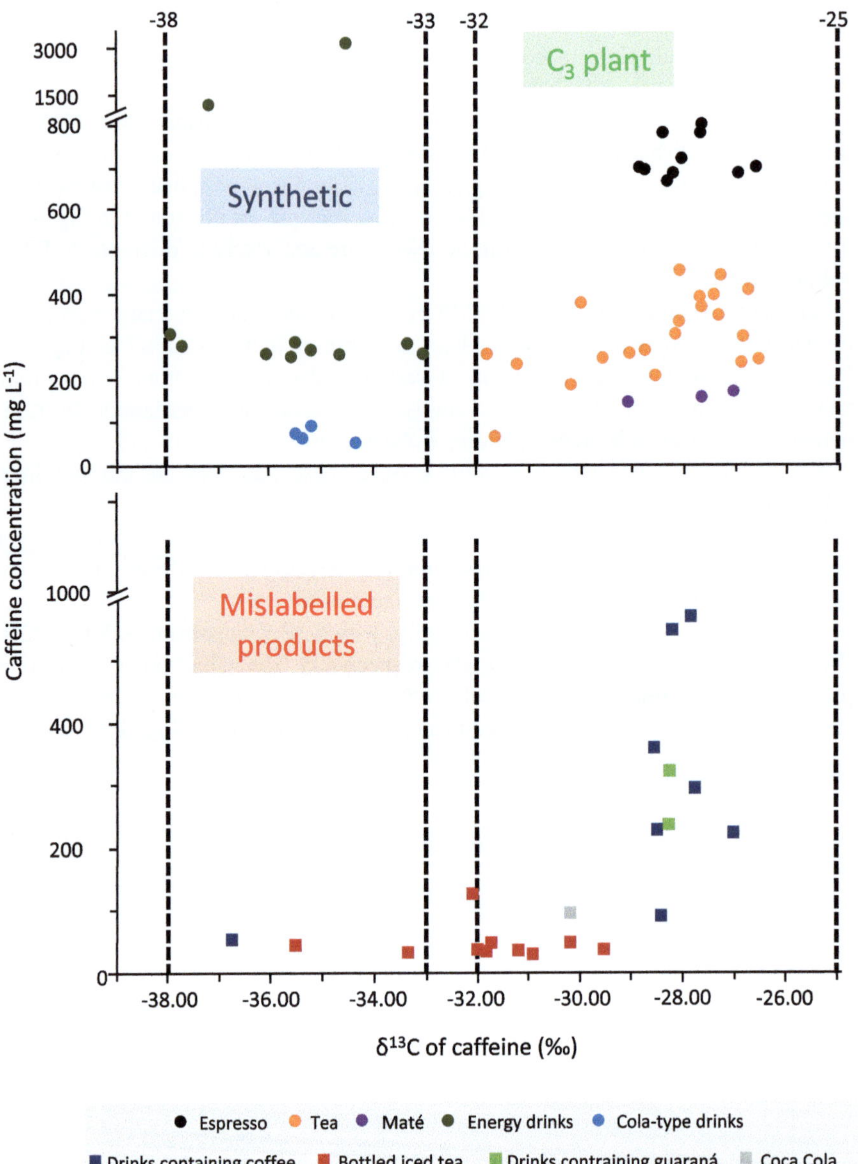

Fig. 4.3 Example for applying isotope analysis to differentiate the natural and synthetic caffeine in various beverages and drinks. This differentiation is used to identify mislabelled products (adapted and modified from Zhang et al. 2012)

As a final example for isotope analyses applied in food characterization, honey sweetening by adding extra sugar can be identified by isotope analysis of monosaccharides (see Fig. 4.4). Sweetening food products such as fruit juice or honey by boosting sugar content with cheap corn syrup derived from the C4 plant maize

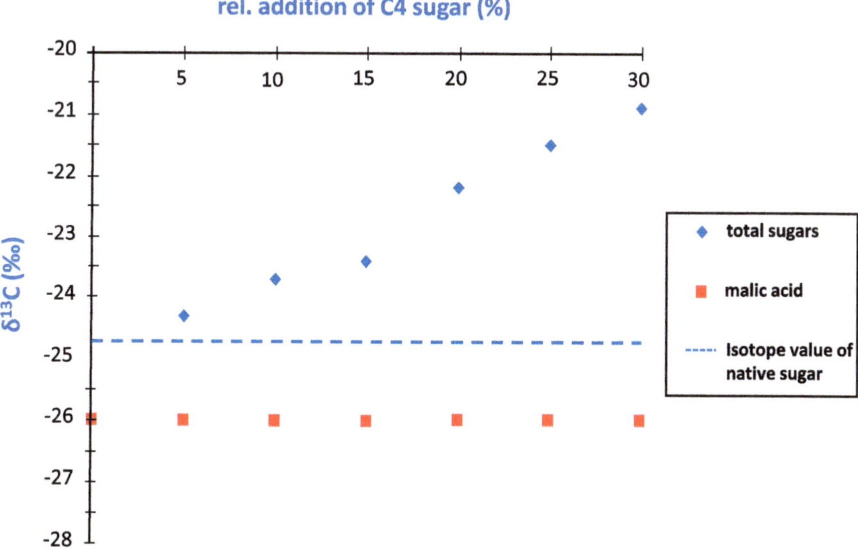

Fig. 4.4 The carbon isotopic shift of honey sugars after adding extra sugar from corn syrup (adapted and modified from Jamin et al. 1997)

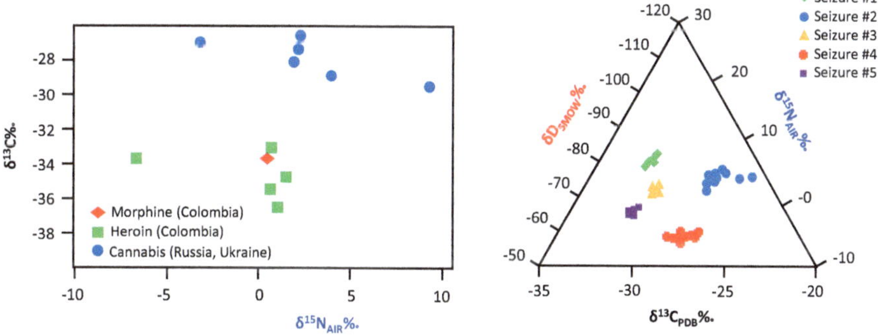

Fig. 4.5 Examples of isotopic characterization of drugs, a two-dimensional separation (carbon-nitrogen) of various drugs from different regions and a three-dimensional discrimination (carbon-nitrogen-hydrogen) of different seizures of ecstasy (adopted and modified after Carter et al. 2002, 2005)

is common. Absolute isotopic values are not used to check the authenticity of honey or related products; instead, the isotopic comparison of biogenetically related constituents like L-malic acid and corresponding monosaccharides (sugars) is used.

The fingerprinting approach is certainly not restricted to the food sector but to related fields. One example is illustrated in Fig. 4.5 for drugs. Here, the geographical origin of different types of drugs can be distinguished by the two-dimensional correlation of nitrogen and carbon isotopes. A second example of application is related to forensic investigations. The question of whether different fractions of heroin confiscated by the police or tablets of synthetic drug (MDMA, ecstasy) belong to the same production batches was answered by multiple isotope analyses. Either a two-dimensional or a three-dimensional correlation of nitrogen and carbon, as well as hydrogen isotope ratios, was suitable for such differentiation.

Furthermore, CSIA can also be applied in sports, e.g., to detect doping cases. A huge challenge is the discrimination of artificial uptake against the individual natural level of performance-enhancing substances such as testosterone. From a chemical point of view, there are no differences in the molecular structure of synthetic and biogenetically produced testosterone. Nevertheless, similar to uncovering artificial sweetening of food, the isotopic composition is the clue for unambiguous identification of the forbidden uptake of additional testosterone, thus doping. There is only one problem: testosterone underlies biochemical transformations, and consequently, direct isotope analyses of testosterone are not optimal. Alternatively, metabolites bearing the same isotopic composition are analyzed. Instead of testosterone, the metabolites androstanediols are often measured. A reliable reference is needed to detect the effect of the uptake of artificial, isotopically different testosterone. Such a reference compound can be another natural steroid whose carbon isotope ratio could not be altered through the uptake of exogenous testosterone, e.g., pregnanediol. An example of the effect on the isotopic composition of these metabolites after an oral uptake of synthetic testosterone is given in Fig. 4.6.

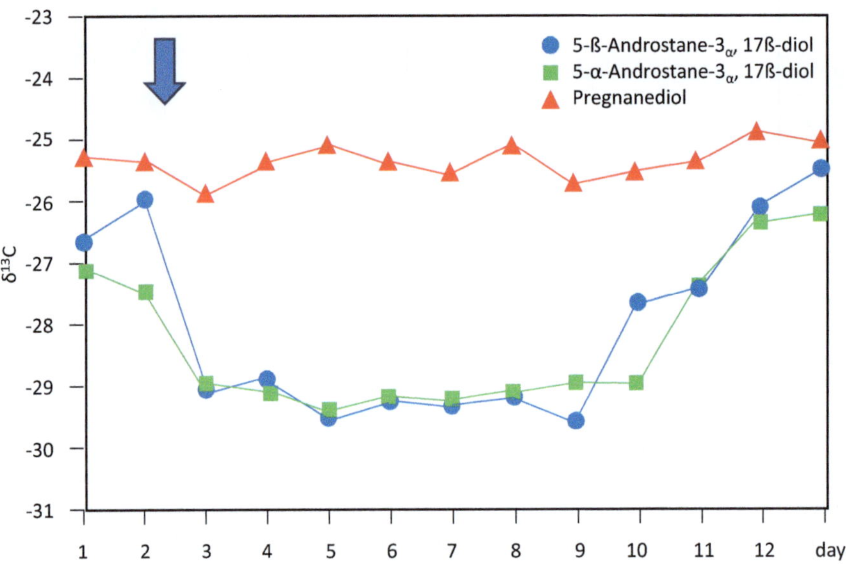

Fig. 4.6 Carbon isotopic shifts of testosterone metabolites in contrast to the reference steroid pregnanediol, measured in urine after uptake of synthetic testosterone at the end of the second day (adopted and modified after Shakelton et al. 1997)

> **General Note**
> Isotopic differences in natural products and synthetic substances are used in various analytical fields, apart from Organic Geochemistry, as a forensic tool for proving inter alia authenticity, regional origin, or doping.

References

Calderone G, Guillou C, Naulet N (2003) Utilisation réglementaire des analyses isotopiques des aliments. l'actualité chimique 20:22–24

Carter JF, Titterton EL, Murray M, Sleeman R (2002) Isotopic characterization of 3,4-methylenedoxyamphetamine and 3,4-methylenedoxymethylamphetamine (ecstasy). Analyst 127:830–833

Carter JF, Sleeman R, Hill JC, Idoine F, Titterton EL (2005) Isotope ratio mass spectrometry as a tool for forensic investigation (example from recent studies). Sci Justice 3:141–149

Ehleringer JR, Thure EC, West JB (2007) Forensic science applications of stable isotope ratio analysis. In: Blackledge RD (ed) Forensic analysis on the cutting edge: New methods for trace evidence analysis. Wiley, New York, pp 399–422

Jamin E, Gonzalez J, Remaud G, Naulet N, Martin GG, Weber D, Rossmann A, Schmidt HL (1997) Improved detection of sugar addition to apple juices and concentrates using internal standard ^{13}C IRMS. Anal Chim Acta 347:359–368

Jochmann MA, Steinmann D, Manuel S, Schmidt TC (2009) Flow injection analysis-isotope ratio mass spectrometry for bulk carbon stable isotope analysis of alcoholic beverages. J Agric Food Chem 57:10489–10496

Shakelton CHL, Phillips A, Chang T, Li Y (1997) Confirming testosterone administration by isotope ratio mass spectrometric analyses of urinary androstanediols. Steroids 62:379–387

Zhang L, Kujawinski DM, Federherr E, Schmidt TC, Jochmann MA (2012) Caffeine in your drink: Natural or synthetic? Anal Chem 84:2805–2810

Further Reading

Benson S, Lennard CL, Maynard P, Roux C (2006) Forensic applications of isotope ratio mass spectrometry—A review. Forensic Sci Int 157:1–22

Chapter 5
Fossil Matter Correlation

Outlook

A step in determining linkages between different fossil materials using CSIA is described here. This includes the correlation of oil/oil systems, as well as the correlation between oil and potential source rocks. These correlations commonly use biomarker analysis of, e.g., hopanes and steranes. However, isotope analysis can provide a substantial improvement in such correlations.

A general question in the Organic Geochemistry of fossil matter is related to identifying genetic linkages of different pools or types of material. One major field is the correlation of two oil reservoirs and whether they are genetically akin. This question is studied by an oil/oil or an oil/source rock correlation primarily based on biomarker analyses using the fingerprint of specific constituents, dominantly hopanes and steranes. The individual composition of these compounds often allows a doubtless assignment. However, isotope analysis provides valuable complementary insights in cases where this approach is unsuccessful (Fig. 5.1).

However, in complex cases, isotope analysis of individual oil or source components allows a deeper insight into genetic linkages. This is related to the isotopic variation due to different processes such as biosynthetic conditions, diagenesis or biotic degradation affecting the individual fossil matter. The differences do not differ highly. Therefore, a very sensitive and thorough analysis is needed. Further, the isotopic analysis is restricted to those components that fit the analytical requirements, such as suitable abundance and successful chromatographic separation. Therefore, n-alkanes are often used as proxies.

Figure 5.2 highlights the variation of oil isotope fingerprints for light hydrocarbons (cyclic and acyclic aliphatics and aromatics). Here, different oils have been characterized as a base for identifying reservoirs filled from mixed origins.

© The Author(s), under exclusive license to Springer Nature Switzerland AG 2024
J. Schwarzbauer, B. Jovančićević, *Isotopes in Organic Geochemistry*, Fundamentals in Organic Geochemistry, https://doi.org/10.1007/978-3-031-69304-5_5

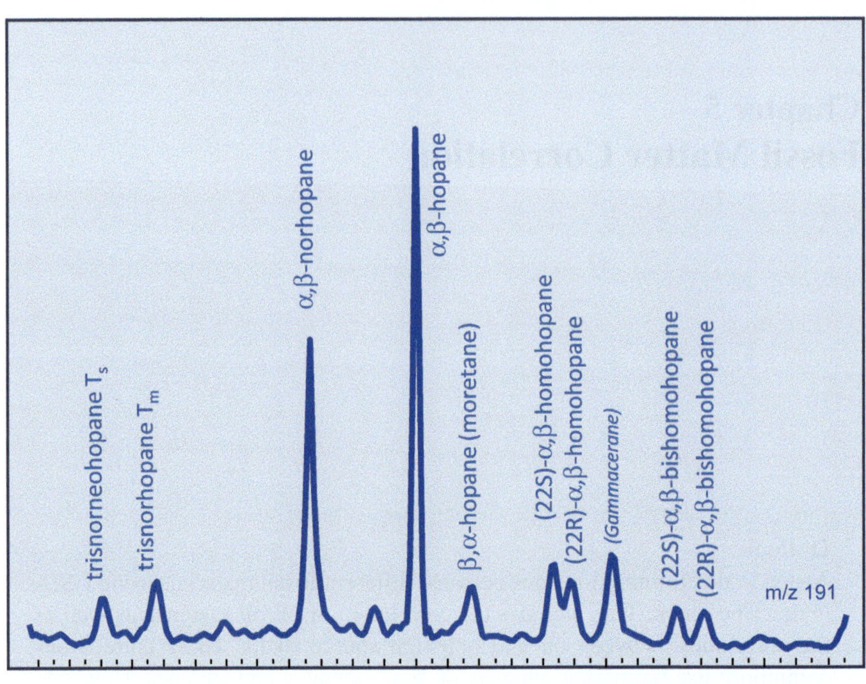

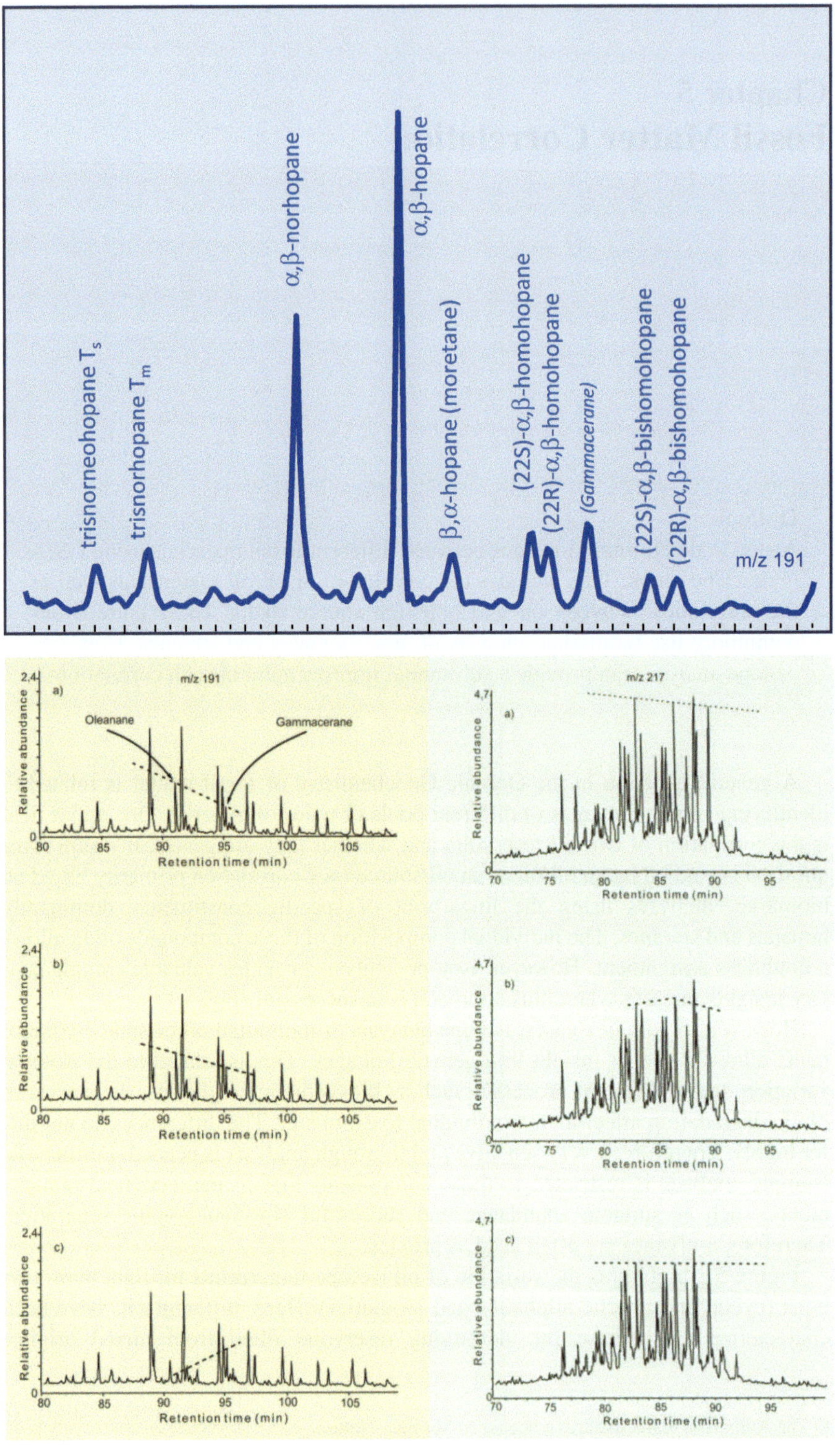

Fig. 5.1 An example of a characteristic pattern of hopanes in a source rock (upper part, illustrated by the m/z 191 ion chromatogram) and an example of an oil/oil correlation on three oils (**a–c**) based on fingerprinting of hopanes and steranes (lower part, illustrated by the m/z 191 and 217 ion chromatograms) (adapted and modified after Jovančićević et al. 1998)

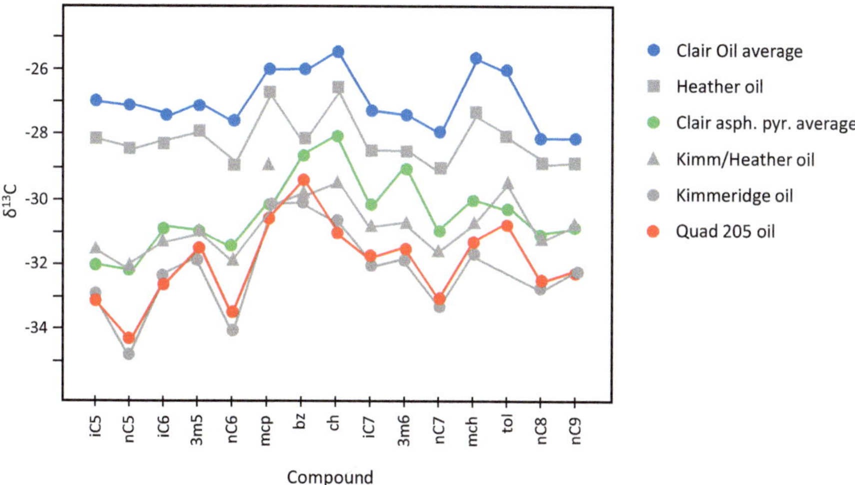

Fig. 5.2 Variations in isotopic compositions of light hydrocarbons (in the gas chromatographic range from *iso*-pentane, iC_5 up to *n*-nonane nC_9) for different oil samples (adapted from and modified after Ronney et al. 1998)

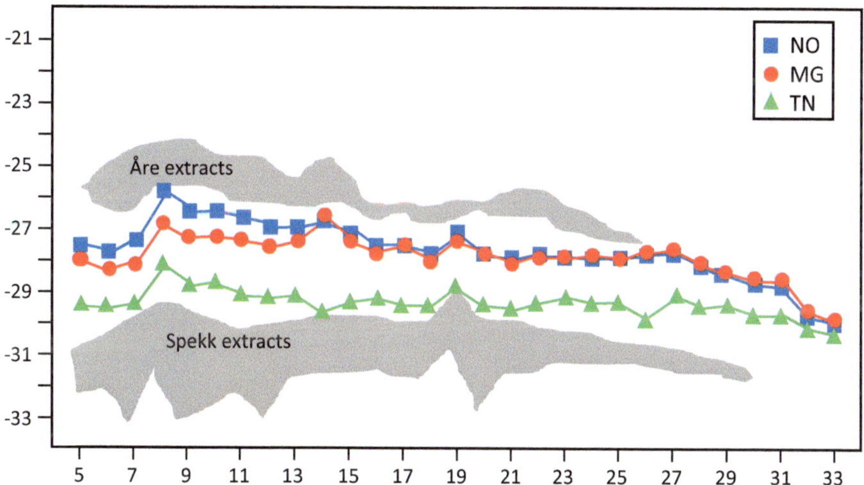

Fig. 5.3 The carbon isotopic signatures of *n*-alkanes from three different Norwegian oils (NO, MG, TN). Not only an apparent similarity between the two oils (NO, MG) is evident, but also their linkage to different potential source rocks, the Åre and Spekk formation (adapted from and modified after Odden et al. 2002)

Overall, the differences of only a few per mill values ($\delta^{13}C$) clearly demonstrate the need for accurate measurements and data interpretation.

An example of successfully applying the carbon isotope analysis approach in correlation studies is illustrated in Fig. 5.3. Here, potential source rocks are intensively isotopically characterized. The information is used not only to differentiate oils from different reservoirs but also to appoint them to the different source rocks. However, it is also evident that the differences are not too high again, meaning that careful interpretation is needed.

References

Jovančićević BS, Polic PS, Vitorovic DK (1998) Organic geochemical investigation of crude oils—the Southeastern part of the Pannonian Basin in Yugoslavia. J Serb Chem Soc 63:397–418
Odden W, Barth T, Talbot MR (2002) Compound-specific carbon isotope analysis of natural and artificially generated hydrocarbons in source rocks and petroleum fluids from offshore Mid-Norway. Org Geochem 33:47–65
Rooney MA, Vuletich AK, Griffith CE (1998) Compound-specific isotope analysis as a tool for characterizing mixed oils: an example from the West of Shetlands area. Org Geochem 29:241–254

Chapter 6
Biogenic vs. Thermogenic Gas Emissions

Outlook
CSIA is also used to differentiate the major natural methane production sources. The isotopic signature of thermogenic methane is much heavier compared to methane produced by microorganisms. This is described in more detail in this chapter by a few examples.

A further application for CSIA in Organic Geochemistry is related to methane and its non-technical production and origin. Generally, natural methane is derived from two major sources: the biological production by dysaerobic microorganisms under more anoxic conditions and the catagenetic formation under higher pressure and temperatures in the deeper sedimentary systems. The corresponding reactions differ significantly. Biogenic synthesis uses complex biochemical systems consisting of enzyme-mediated reactions. The thermogenic formation of methane is based on the abiotic cleavage of longer carbon chains, hence triggered by thermodynamic stability. As described more theoretically in Chap. 2, different reaction pathways affect the isotopic composition of the products. This is also valid for the two different types of methane production.

A notable aspect of methane isotopic characterization is the application of two-dimensional correlation by combining carbon and hydrogen isotope ratios. This allows a highly specific attribution to different pools of natural methane, as illustrated schematically in Fig. 6.1, considering the different natural methane pools comprehensively.

However, Organic Geochemistry focuses mainly on differentiating biogenic from thermogenic methane. Some examples are the following.

A vital aspect of monitoring and managing environmental risks is characterizing methane sources in shallow groundwater and identifying the contribution of methane accompanying bituminous shale, coal or oil deposits. Regularly, shallow source

J. Schwarzbauer, B. Jovančićević, *Isotopes in Organic Geochemistry*,
Fundamentals in Organic Geochemistry,
https://doi.org/10.1007/978-3-031-69304-5_6

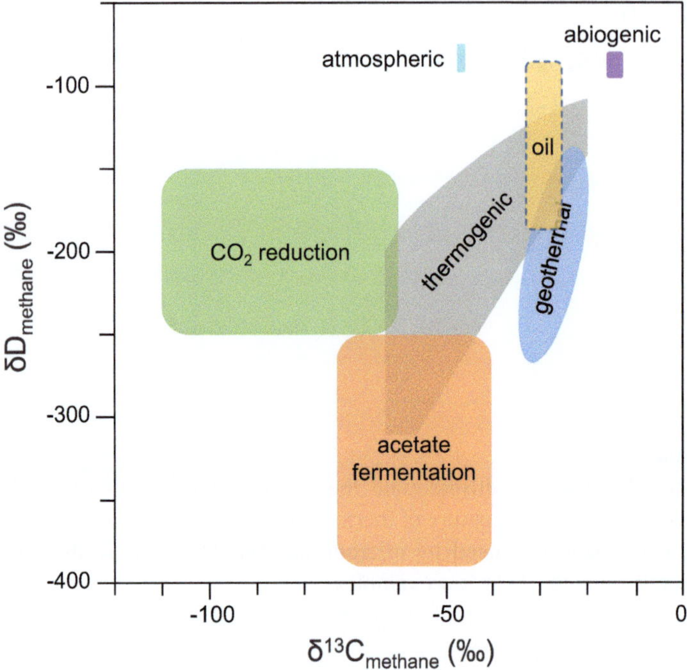

Fig. 6.1 Principal differentiation of methane origin and occurrence according to the two-dimensional correlation of carbon and hydrogen isotope characteristics

methane is of biogenic origin, appearing with other specific gases such as CO_2, H_2S, H_2O, NH_3 or N_2. As mentioned, it differs significantly from methane produced in late catagenesis or metagenesis by the carbon isotopic composition, δ13CPDB. In contrast to an isotopic two-dimensional correlation, the gas composition with respect to the early members of the homolog series of n-alkanes, whose starting compound is methane, can be used for this approach. Biogenic gas is composed more or less exclusively of methane, whereas petrogenic gases also exhibit relevant amounts of ethane, butane and higher homologs. Hence, the carbon isotope characteristic can be attributed to the gas composition, revealing a so-called 'Bernhard plot.' This has been used by Györe et al. (2018) for characterizing coalbed methane from central Scotland and coalmine methane from central England as potential sources for shallow gas pools (see Fig. 6.2). Here, isotopic values for δ¹³CPDB were determined in the range from −39.5 to −51.1‰. These are in the range that is characteristic of gases of thermogenic origin and differ from the values of most gases from shallow sources in the UK.

A more general example represents the data summarized in Fig. 6.3. Here, coal bed-derived methane is compared with further thermogenic gases (e.g., North Sea gas reservoirs) and methane derived from biogenic sources such as landfills. As already introduced, it is based on the two-dimensional plot of hydrogen and carbon isotope values.

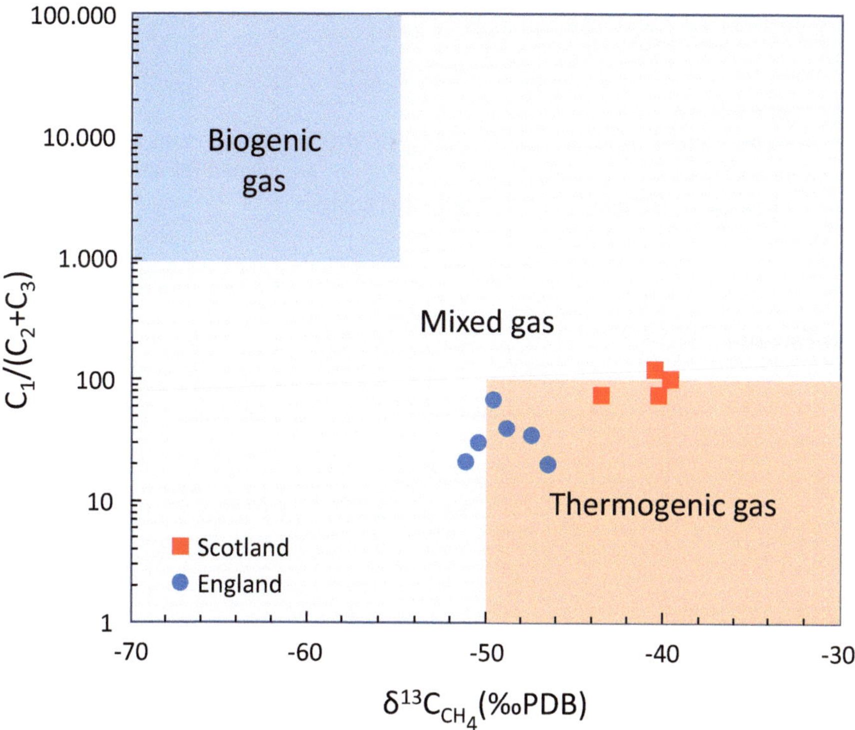

Fig. 6.2 The 'Bernard plot' correlating molecular and carbon isotope composition of coal-derived gases from the UK. Both England and Scotland gases are clearly of thermogenic origin (adapted and modified from Györe et al. (2018))

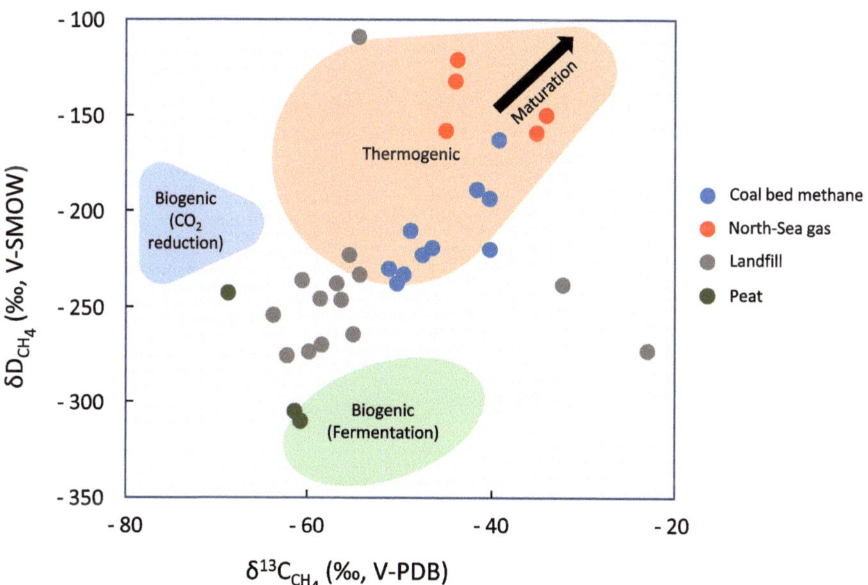

Fig. 6.3 Two-dimensional correlation of hydrogen and carbon isotopic compositions of methane from various pools and the linkage to principal sources (adapted and modified from Györe et al. (2018) and data therein)

> **General Note**
> Isotopic composition is a key aspect of differentiating natural methane sources. However, two-dimensional correlations are needed for an unambiguous linkage of methane pools to individual sources.

Reference

Györe D, McKavney R, Gilfillan SMV, Stuart FM (2018) Fingerprinting coal-derived gases from the UK. Chem Geol 480:75–85

Chapter 7
Paleoenvironmental Conditions

Outlook
One of the main applications of isotope analyses in Organic Geochemistry is related to the reconstruction and characterization of the paleoenvironment and paleoclimate. The biogeochemical cycles, in particular, influence the isotopic composition of fossil organic material, which allows the inspection of their characteristics via isotope analysis of sedimentary archives.

Isotope analysis opens the possibility of gaining information about the paleoenvironment, past natural conditions and biological composition. Here, the influence of processes and conditions on photosynthesis is a main clue. Changes influencing the biological reaction conditions also affect the resulting isotopic composition of organic matter.

First principal studies on paleoenvironmental reconstructions (e.g. Freeman et al. 1990; Hayes et al. 1990) figured out these interrelations in a more detailed way such as focussing on carbon isotope signals in sedimentary rocks deriving from biogeochemical processes of different communities such as algae and bacteria. Here they linked certain paleoenvironmental conditions and the global carbon cycle with the microbial community e.g. of lacustrine areas. Further on, organic matter in sedimentary rocks is suggested to originate from various sources, not only from ancient primary producers but also from consumers and secondary producers. These general species can be differentiated by varying $\delta^{13}C$-values that derive from various biogeochemical processes and mechanisms as well as biosynthetic reactions involving the metabolism of the communities. In more detail, the $\delta^{13}C$-values hereby can tell whether the organisms were, e.g., heterotrophic, chemoautrophic or ammonia-oxidizing. During the preservation and diagenesis stage of a sedimentary rock, the typical $\delta^{13}C$ fingerprints are only minorly influenced. For primary organic matter, porphyrin can used as proxy by following its diagenetic products, pristane and

J. Schwarzbauer, B. Jovančićević, *Isotopes in Organic Geochemistry*, Fundamentals in Organic Geochemistry, https://doi.org/10.1007/978-3-031-69304-5_7

phytane. This is a commonly used tool in paleoenvironmental reconstruction due to their distinct carbon isotopic compositions of –25.4‰ (pristane) and –31.8‰ (phytane), showing a strong correlation with the corresponding fingerprint of geoporphyrins. The studies further reveal that the isotopic variation within the sedimentary rock samples is more significant than differences in total organic carbon (Fig. 7.1).

One of the first studies exemplifying the potential of compound-specific isotope analysis to gain detailed information on the biogenic sources of organic matter in sedimentary records was by Rieley et al. (1991), which examined sediments in Ellesmere Lake (UK). They analyzed the carbon isotope ratios ($\delta^{13}C$) of specific n-alkanes from leave waxes derived from trees around the lake and compared their results with corresponding data from the lake sediments. As a result, they intended to differentiate individual biogenic sources of sedimentary carbon.

As analyzed in the leaf waxes, the isotopic values of n-alkanes ranging from C_{25} to C_{33} differed due to genetically derived inter-specific variation from $-30.1‰$ to $-38.7‰$ (see Fig. 7.2). Short-chain n-alkane homologs, derived from autochthonous sources, dominantly from algae, demonstrated negative values ranging from $-22.7‰$ to $-23.8‰$. However, the $\delta^{13}C$ values for C_{25} to C_{33} n-alkanes in sediment samples vary from $-30.1‰$ to $-35.9‰$, clearly pointing to a significant input of leaf material from lakeside trees.

This very early example of a successful organic geochemical application of isotope analyses points to the potential use of CSIA to reconstruct ancient biogeochemical processes.

A second example demonstrates such application, pointing to the effect of paleoclimatic changes on the carbon isotope composition of specific biomarker compounds. Climate changes are characterized inter alia by temperature variations that affect the isotope composition of biosynthetic products in a complex manner. The study of Yamada and Ishiwatari (1999) focused on n-C_{29} and n-C_{31} alkanes and their carbon isotopic composition in a core reflecting a sedimentation period of the last 85,000 years (see Fig. 7.3). As mentioned, these long-chain n-alkanes reflect the organic matter from higher land plants due to their diagenetic origin from corresponding carboxylic acids (with one more carbon atom), which are specific

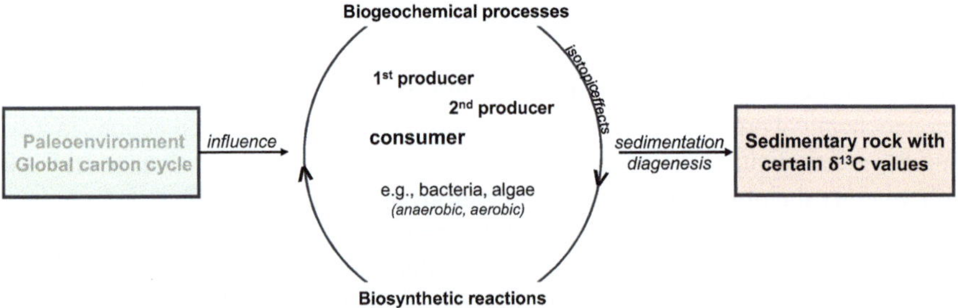

Fig. 7.1 Principal aspects in linking isotopic characteristics of individual biogenic compounds with their corresponding chemofossils (adapted from and modified after Freeman et al. 1990 and Hayes et al. 1990)

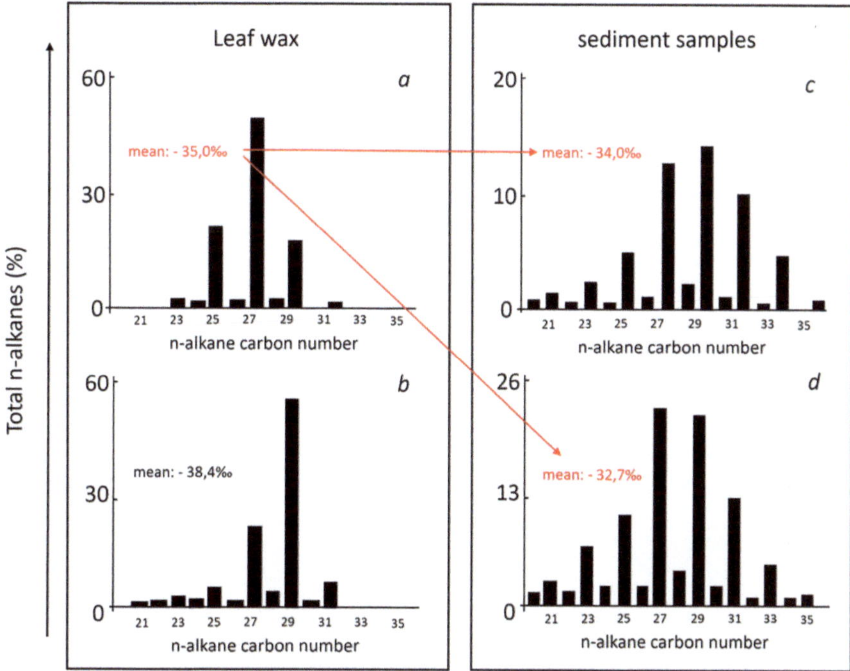

Fig. 7.2 An example of how the source of sedimentary input can be differentiated based on the carbon isotope ratios. For this purpose, two leaf waxes (*Salix alba* **a** and *Aesculus hippocastanum* **b**) and two sediment samples from 0 to 5 and 10 to 15 cm (**c, d**) were examined. The $\delta^{13}C$ values of the leaf waxes in the range of C_{25}–C_{29} closely resemble those of the sediment samples. The $\delta^{13}C$ value of *Salix* is much closer to those observed in the sediment samples. Hence, *Salix* can be considered a significant contributor (adapted from and modified after Rieley et al. 1991)

constituents in cuticular waxes of higher land plants. Some main shifts in isotopic composition were evident within the core, which correlated with known climate changes. During a glacial phase, around 10,000 years ago, the $\delta^{13}C$ values of the *n*-alkanes shifted to more negative values, reflecting a lighter isotopic ratio. The shift can be attributed to variations in terrestrial primary productivity and/or changes in the C3/C4 plant ratio. This ratio is highly temperature-sensitive and applicable for climate reconstruction. Generally, during colder climates, there is an overall decrease in biomass production, but the relation of C3 to C4 plant growth is relatively enhanced because C4 plants favor higher temperatures.

Noteworthy, this study also used a bulk isotope parameter, the $^{18}O/^{16}O$ ratios ($\delta^{18}O$), to reconstruct warmer or colder paleoclimatic conditions or even glacial episodes (see Fig. 7.2). This approach is based on the already described isotopic shift effect of water evaporation and precipitation (see Chap. 2), which are also strongly linked to climate. During a warmer climate, the ocean water is enriched in ^{18}O and depleted in ^{16}O, whereas during a colder climate, the ocean water is enriched in ^{16}O and depleted in ^{18}O. As a secondary consequence, ^{16}O is preferably

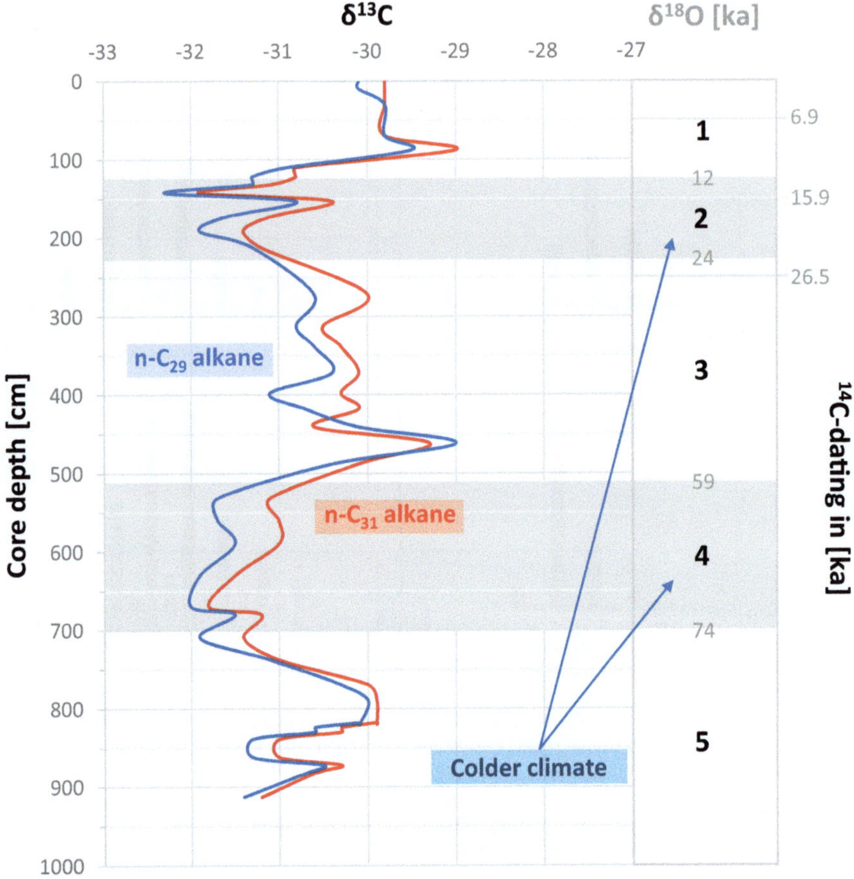

Fig. 7.3 The relation of isotopic shifts of specific *n*-alkanes and related climate changes (adapted from and modified after Yamada and Ishiwatari 1999)

incorporated into the ice during a glacial phase. Hence, the discussed study shows that bulk $\delta^{18}O$ values can be used as an indicator of climate conditions.

A very similar approach has been reported by Tanner et al. (2007). Here, the isotopic differences in C3 and C4 plants are used to reconstruct a change in plant community and to confirm a change in sea level in the late Holocene. The analyses were applied to salt marsh sediments from the state of Maine. By analysis of two long-chain odd-numbered *n*-alkanes, the contribution and variations of higher land plants were followed. As exemplified in Fig. 7.4, for $n\text{-}C_{27}$, the concentration data do not reflect the environmental changes, whereas the corresponding carbon isotope values point to significant paleoenvironmental changes. It was evident that the sea level was rising during the late Holocene, as indicated by a change from a C3 plant-dominated marshland towards an ecosystem with a mixed C3 and C4 community. Here, isotopic analyses represent a more suitable tool for reconstructing

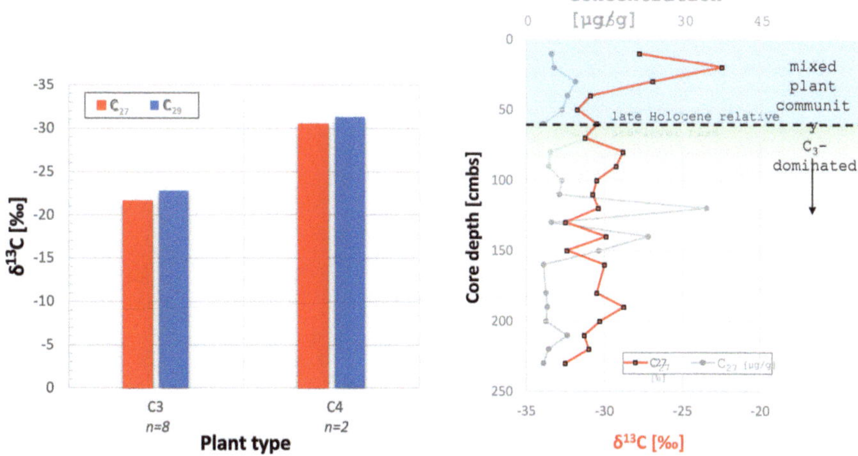

Fig. 7.4 Contrasting the specificity of quantitative data and isotope values for paleoenvironmental reconstructions exemplified for n-C$_{27}$ shifts in a sedimentary record (adapted from and modified after Tanner et al. 2007)

paleoenvironments than quantitative analysis. However, for both approaches, selecting appropriate biomarkers is the essential precondition.

The more specific the relation between biomarkers and plant species, the more detailed paleoenvironmental changes can be traced. This is reflected in a study by Hyun et al. (2017), which combined bulk carbon isotopic data ($\delta^{13}C_{org}$), already mentioned long-chain n-alkanes and their compound-specific carbon isotope ratios ($\delta^{13}C_{ALK}$), for reconstructing the paleoclimatic conditions and paleovegetation changes over the last 35 millennia.

Three differing sections and corresponding time periods could be distinguished based on typical geochemical bulk data such as TOC, TN and the corresponding isotopic data (δ^{13}Corg and δ^{15}N), of a sedimentary archive derived from a Korean maar lake. Comparing the three sections (as given in Fig. 7.5), the $\delta^{13}C_{ALK}$ values provided evidence for an isotopic shift from a predominance of C4-plants (with some mixture of C3-plants) towards exclusive C3-plant vegetation. This conclusion is based on the already discussed differences in isotope ratios of C4-derived n-alkanes with lower values of ca. -18 to $-26‰$ and C3-derived n-alkanes showing lighter signals ranging from ca. -31 to $-39‰$. In this study, the described changes in paleovegetation were also supported by the total n-alkane concentrations. Here, the results show a Holocene increase and a glacial decrease with an intermediate concentration in the transition period from Glacial to Holocene times.

However, organic-geochemical isotope analyses are certainly not restricted to n-alkanes. In the study of Upadhayay et al. (2024), dicarboxylic acids were investigated for their potential use as specific indicators. The analyzed dicarboxylic acids are specific for suberin, an important biopolymer of plant roots. Long-chain

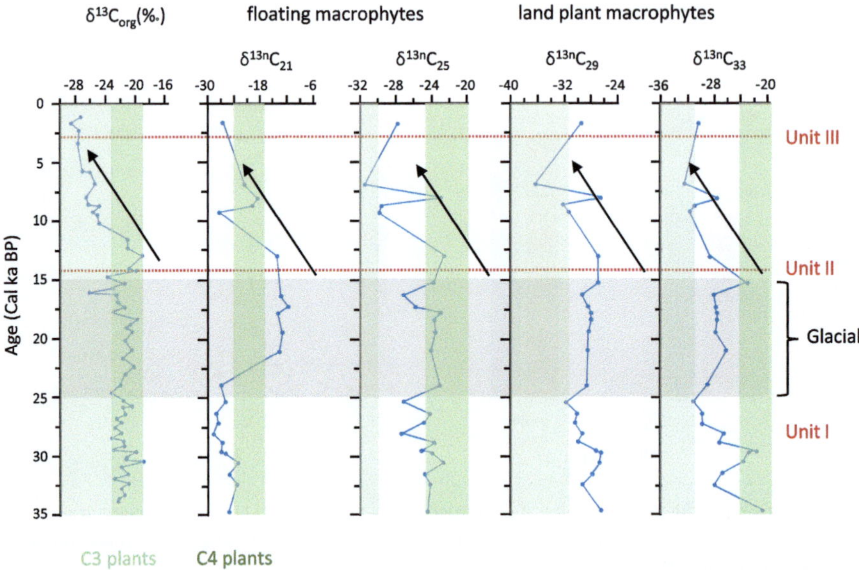

Fig. 7.5 The carbon isotope record of individual *n*-alkanes for differentiation of C3 and C4 plants in the paleoenvironment and its implication for paleovegetation changes as a result of climate shifts (adapted from and modified after Hyun et al. 2017)

dicarboxylic acids are very stable in soil due to their strong adsorption on clay minerals, and their capacity to survive long-range transport in river systems.

In this study, the characteristics of the indicative long-chain dicarboxylic acids (C_{18}-C_{26}) and their specific $\delta^{13}C$ signature have been used to follow changes in land usage over time and to indicate heavy rainfall periods. Analyses of four different soil types representing woodland, pasture, arable and stream banks were performed. A change of rainfall intensity was followed in one winter period, distinguishing early winter (EW) and late winter (LW) time, in which the late winter period has a higher rainfall intensity and abundance.

The results showed that the highest amount of the dicarboxylic acids was detected in woodland, followed by pasture, arable and stream banks, pointing to a more intensive degradation of these biomarkers with intense agricultural land use. However, following dicarboxylic acids and their corresponding $\delta^{13}C$ signatures (see Fig. 7.6) in time-integrated riverine suspended particulate matter (SPM) samples, a linkage to rainfall events can be established due to erosion. In the early winter, eroding stream banks were the dominant source of riverine suspended particulate matter. In contrast, the arable land was the dominant contributor to SPM in the later winter, when heavy rainfalls occurred. In conclusion, isotopic characteristics of root-derived dicarboxylic acids can act as indicators for quantifying the effects of heavy rainfalls on soil mobilization.

Compound-specific isotope analyses often combine with distinct analytical methods or approaches to reveal comprehensive paleoinformation. This will be illustrated by discussing a recent study by Weniger and Schwarzbauer (2024) in the following.

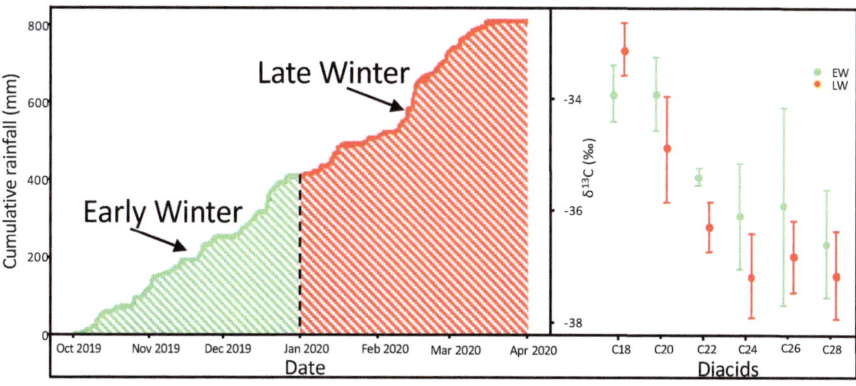

Fig. 7.6 Rainfall characteristics in two winter periods and the corresponding carbon isotope signals of specific dicarboxylic acids differentiating these two time periods (adapted from and modified after Upadhayay et al. 2024)

In this study, not only extractable *n*-alkanes, as well-established biomarkers, were measured but also *n*-fatty acids were also considered. Both substance groups have been measured in paleozoic coals of low to medium thermal maturity to follow signals with only low influence from diagenetic or catagenetic alterations (for more detailed information, check Schwarzbauer and Jovančićević 2016). The extractable components (free fraction) were analyzed as usual, while the bound fraction was also investigated. Since fatty acids are often linked to organic matter by ester bonds, an alkaline hydrolysis approach has been applied to cleave these bonds. As illustrated in Fig. 7.7a, the pattern of the homolog series of both free and bound fatty acids differ significantly. The free fatty acids are characterized by a substantial contribution of long-chain compounds ($>C_{20}$), indicating an elevated contribution of cuticular wax-derived matter. These signals (particularly fatty acids with $>C_{26}$) are observed in higher land plant material. The common fatty acids with C_{16} and C_{18} were dominant in the bound fractions, which can be attributed generally to natural fats and oils (triglycerides). As a second observation, the even-odd predominance of the free fatty acids is lower than that of the bound compounds. This is reflected by the carbon preference indices (CPI), the differences of which are illustrated in Fig. 7.7b.

Compound-specific carbon isotope analyses (CSIA) were applied to all substance groups. The fatty acids (after derivatization to methyl esters) have been detected with a sufficient separation of the individual isomers (see Fig. 7.7c), a critical precondition for unambiguous and reliable detection of isotope values. As a first result, comparing isotope shifts within the homolog series of *n*-alkanes and *n*-carboxylic acids revealed a clear difference. The isotope values within the *n*-alkane series remain more or less constant, whereas a shift toward lighter isotopic composition was evident for the fatty acids series (see Fig. 7.7d). This was valid for all investigated samples, as exemplified in Fig. 7.7e. To interpret these shifts, the biochemical synthesis of fatty acids with different chain lengths has to be considered.

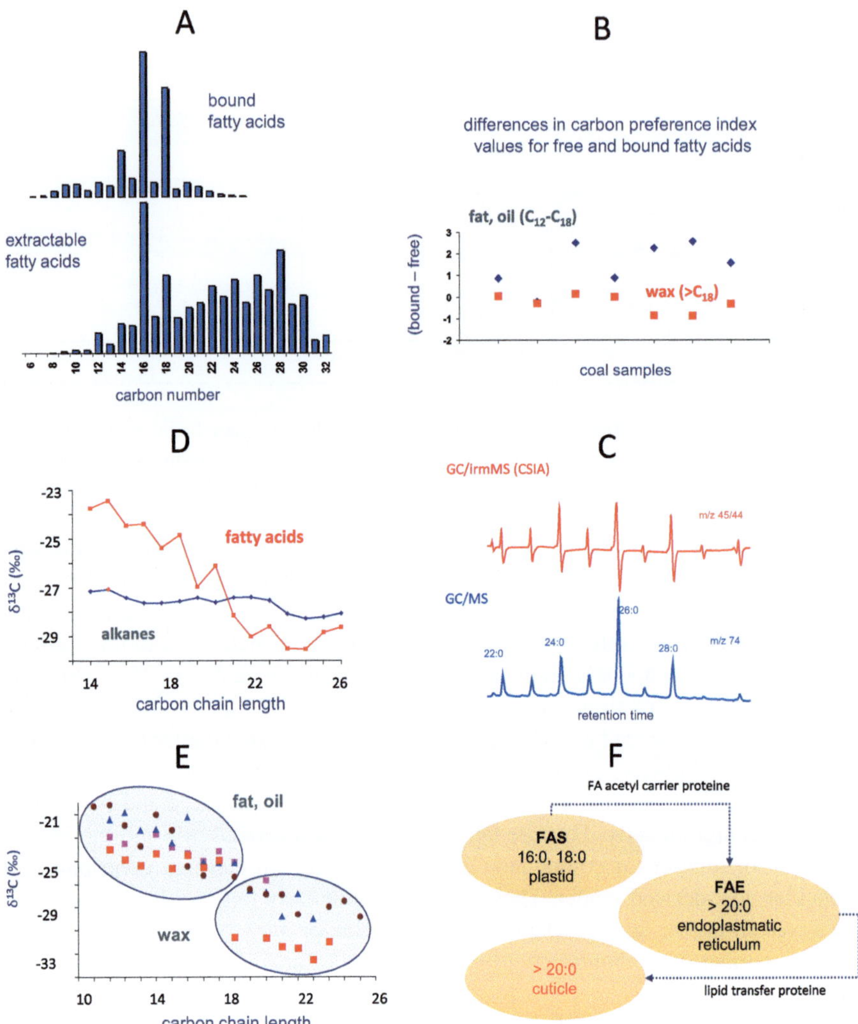

Fig. 7.7 Summary of main results from analyses of *n*-alkanes and fatty acids in selected Paleozoic coals. (**a**) Example for the homologs pattern of detected free and bound fatty acids; (**b**) differences in carbon preference values for free and bound fatty acids; (**c**) the 'swing' from compound-specific carbon isotope analysis and the corresponding ion chromatogram as proof for successful detection and isotopic analysis of the fatty acid (methyl esters), (**d**) example for the trends in the carbon isotope composition of *n*-alkanes and corresponding fatty acids; (**e**) four examples for shifts in the carbon isotope composition of fatty acids with different chain lengths; (**f**) Principal pathway of the biochemical fatty acid synthesis concerning different chain lengths (all data are adapted from and modified after Weniger and Schwarzbauer 2024)

Fatty acids with C_{16} and C_{18} chain lengths (palmitic and stearic acids) as main constituents in triglycerides (natural fats and oils) are synthesized in the plastids, whereas longer chain lengths are formed after an intracellular transfer in the endoplasmatic reticulum. Finally, these wax-related fatty acids are then transferred to the cuticle. Therefore, slightly different biochemical synthesis pathways, intra- and extracellular transport, lead to different isotopic characteristics for these two fatty acid groups. The difference is still visible in Paleozoic coals—if isotopic measurements are combined with bound residue analyses.

> **General Note**
> Compound-specific isotope analysis is a valuable tool for gaining information on paleoenvironmental and paleoclimate conditions. The complementary application of further analytical approaches is often beneficial for obtaining comprehensive information.

References

Freeman KH, Hayes JM, Trendel JM, Albrecht P (1990) Evidence from carbon isotope measurements for diverse origins of sedimentary hydrocarbons. Nature 343:254–256

Hayes JM, Freeman KH, Popp BN, Hoham CH (1990) Compound-specific isotopic analyses: A novel tool for reconstruction of ancient biogeochemical processes. Org Geochem 16:1115–1128

Hyun S, Shin KH, Lee SC, Chang SW, Nam SI (2017) Terrestrial *n*-alkanes and their carbon isotope records from the Hanon paleo-maar sediment, Jeju Island, Korea: Implications for paleoclimate and paleovegetation over the last 35 kyrs. Quatern Int 441:89–100

Rieley G, Collier R, Jones D, Eglinton G, Eakin PA, Fallick AE (1991) Sources of sedimentary lipids deduced from stable carbon-isotope analyses of individual compounds. Nature 352:425–427

Schwarzbauer J, Jovančićević B (eds) (2016) Fundamentals in organic geochemistry, vol 2: From biomolecules to chemofossils. Springer, Cham. 160 pp, ISBN 978-3-319-27241-2

Tanner BR, Uhle ME, Kelley JT, Mora CI (2007) C3/C4 variations in salt-marsh sediments: An application of compound specific isotopic analysis of lipid biomarkers to late Holocene paleoenvironmental research. Org Geochem 38:474–484

Upadhayay HR, Joynes A, Collins AL (2024) ^{13}C dicarboxylic acid signatures indicate temporal shifts in catchment sediment sources in response to extreme winter rainfall. Environ Chem Lett 22:499–504

Weniger P, Schwarzbauer J (2024) Organic geochemical and compound specific stable carbon isotope analyses of n-alkanes and saturated carboxylic acids, extracted from late Paleozoic coals. Int J Earth Sci, submitted

Yamada K, Ishiwatari R (1999) Carbon isotopic compositions of long-chain n-alkanes in the Japan Sea sediments: implications for paleoenvironmental changes over the past 85 kyr. Org Geochem 30:367–377

Chapter 8
Isotope Analysis in Environmental Sciences

Outlook

This final chapter explains, by selected examples, the general application fields of CSIA in environmental organic geochemistry. Besides fingerprinting for emission source identification, isotopic shifts also point to biotic degradation processes in natural ecosystems.

Compound-specific isotope analyses are dominantly used in environmental sciences, with two different approaches already introduced. On the one hand, fingerprinting is applied in environmental forensics, focusing on emission sources and pollution origin. Moreover, the highly relevant process of biodegradation of pollutants as one major aspect of their life cycle can be followed and quantified by isotope analysis. Following, some examples of the isotope fingerprint approach are introduced.

Already in 1994, the carbon isotope composition of PAHs has been studied to perform an emission source apportionment (O'Malley et al. 1994). Despite the general origin of PAH from principal sources (petrogenic and pyrogenic ones), a more detailed linkage of PAH contaminations to individual emission sources is often needed. In this study, CSIA has been applied for this task at a very early stage of application of isotope analyses on environmental problems. Hence, firstly, the method's suitability has been tested, e.g., by checking the influence of extraction procedures, transport, biotic degradation, or sunlight exposure to the isotopic composition of main PAHs (see Fig. 8.1, left). As expected, microbial degradation has a significant influence, whereas all other aspects have a negligible impact on carbon isotope composition. Based on this proof, three- to five-ring PAHs from different sampling sites (bay and harbor sediments) and potential (secondary) emission sources (such as road sweeps or sewage) have been measured and tried to link (see Fig. 8.1, right).

J. Schwarzbauer, B. Jovančićević, *Isotopes in Organic Geochemistry*, Fundamentals in Organic Geochemistry, https://doi.org/10.1007/978-3-031-69304-5_8

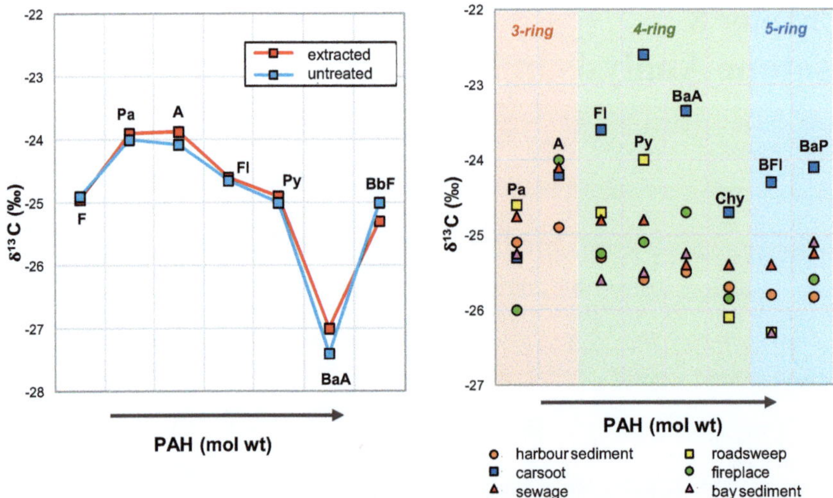

Fig. 8.1 No change of PAH isotopic composition before and after extraction (left) and carbon isotope characteristics of selected PAHs in environmental samples and suspected secondary emission sources (right)—Pa, phenanthrene; A, anthracene; Fl, fluoranthene; Py, pyrene; BaA, benz(a) anthracene; Chy, chrysene; BFl, benzofluoranthenes; BaP, benz(a)pyrene

Yanik et al. (2003) studied the application of CSIA on PCBs for a source appointment approach at the Housatonic River area (Pittsfield, Massachusetts) on a more in-depth level. The principal challenge was to link PCB contamination in biota with a pollution source. For this purpose, duck muscle, liver and egg samples, as well as two grass carp samples, were analyzed. As suspected, the pollution source of the reported discharge, Arochlor 1260, in this area was considered. Besides, biota samples of various technical PCB formulations (Arochlor 1242, 1254, 1260 and 1262) with various chlorine contents were compared in two different ways. Since PCBs occur as a congeneric mixture with various degrees of chlorination and different isomers within one group of homologs, the resulting typical pattern of isomers has been compared by standard GC/MS analysis. Here, a clear technical origin of PCBs in the animals was evident (see Fig. 8.2). However, within one degree of chlorination, the pattern does not differ between the technical mixtures, meaning their differentiation based solely on isomer composition failed.

Applying both congener-specific pattern recognition and carbon isotope analysis on selected congeners, the proposed contamination source (Arochlor 1260) became evident. The deviations between data in biota samples and the technical mixture were low, suggesting a significant correlation was confirmed.

Besides the pattern recognition approach, CSIA has been applied in environmental sciences, especially following a vital process related to organic pollution, namely biotic degradation. As introduced in Chap. 2, the (bio)synthesis and the microbial-assisted degradation of organic compounds systematically affect their isotopic composition.

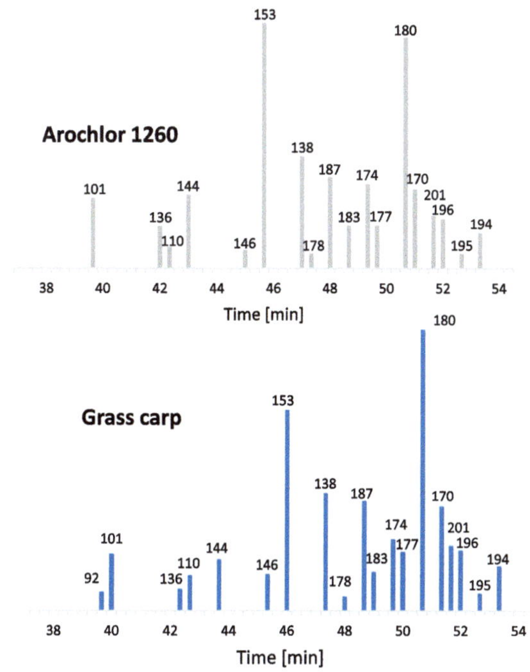

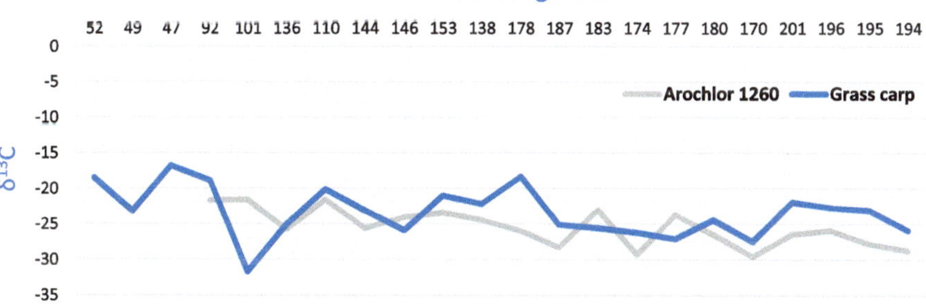

Fig. 8.2 PCB pattern and isotopic differentiation of PCB contamination in biota as compared to the potential emission characteristics (adapted from and modified after Yanik et al. 2003)

This is clearly demonstrated, for example, in the studies of Meckenstock et al. (2004), Morasch et al. (2001) and Griebler et al. (2004). They intended to follow biodegradation in contaminated aquifers under anoxic conditions by observing isotope fractionation, e.g., of toluene. Their studies are based on the kinetic isotope effect during biodegradation quantifiable by the Rayleigh equation, as inserted in Fig. 8.3. Here, C_t/C_0 describes the fraction (f) of the substrate that remains after a certain time, and R_t/R_0 describes the corresponding isotope ratios. As a result, a linear correlation of both parameters can be obtained, as illustrated in Fig. 8.3.

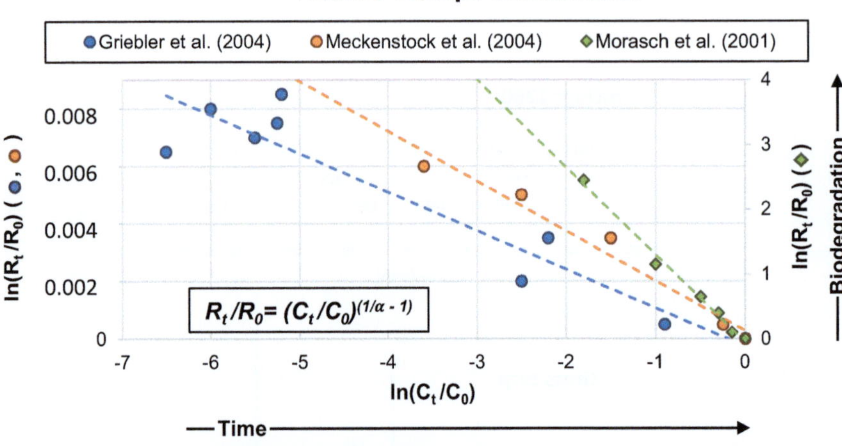

Fig. 8.3 The isotopic characterization of microbial toluene degradation

However, this approach is not limited to carbon isotopes. In the studies of Griebler et al. (2004) and Meckenstock et al. (2004), the stable carbon isotopes were measured, while the study of Morasch et al. (2001) focussed on D/H isotope fractionation. In the case of halogenated contaminants, the halogen isotopes (particularly chlorine) are also helpful. E.g., Sturchio et al. (1998) and Meckenstock et al. (2004) investigated contaminated sites for trichloroethene (TCE). Here, Cl isotope ratios were investigated to evaluate natural attenuation processes in the contaminated aquifers, proving to be a successful tool. A very comprehensive isotope approach was used by following the underground biotic degradation of a similar pollutant, tetrachloroethylene, by complementary analyses of hydrogen (δD), carbon ($\delta^{13}C$) and chlorine ($\delta^{37}Cl$) isotope shifts (Filippini et al. 2018). The shift for $\delta^{37}Cl$ values is illustrated in Fig. 8.4.

Noteworthy, these systematic shifts are also valid for abiotic degradation processes, as shown in Fig. 8.5 for the iron-assisted dehalogenation of trichloroethylene.

The degradation shift of especially carbon isotopic composition is applied intensively in studies related to groundwater contamination by aromatics and halogenated aliphatics. This approach was dominantly used for monitoring natural attenuation, considering bioremediation measures and management approaches for groundwater contamination.

According to Rayleigh, the base for calculating degradation processes in open systems like aquifers is the linear correlation, as illustrated in Fig. 8.1. However, the fractionation factor α or the enrichment factor ε (with $\alpha = \varepsilon + 1$) as the key parameter needs to be determined before any analyses of natural samples. Laboratory experiments usually do this. An overview of some enrichment factors for different

Fig. 8.4 Following the biotic degradation of tetrachloroethylene by the stable chlorine isotopic composition (adapted and modified after Filippini et al. 2018)

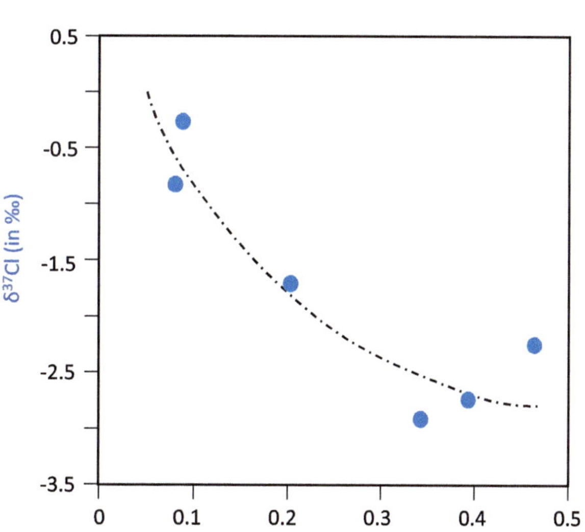

Fig. 8.5 The carbon isotopic shift of abiotic trichloroethylene degradation and its corresponding enrichment factor ε, here exemplified in a direct (exponential) relation between isotope values and degraded proportions (adapted from and modified after Slater et al. 2002)

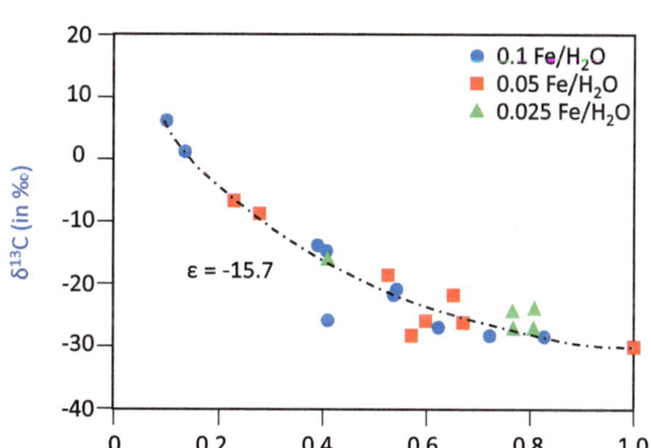

compounds and types of biodegradation, as well as various isotopes exemplifying their variances, are summarized in Table 8.1.

An example of an application of compound-specific isotope analysis CSIA in the field of groundwater contamination is exemplified in Fig. 8.6. A point source pollution by hexachlorocyclohexanes (HCHs, including the isomer γ-HCH, also known as lindane) was monitored directly within the plume and downstream in the contaminated aquifer. A decrease in concentration along the flow direction of the groundwater was measured, but the question arose whether it was the result of

Table 8.1 Examples of enrichment factors determined in laboratory experiments for carbon, hydrogen and chlorine under various biodegradation conditions

Substance group	Conditions	Range of enrichment factors ε		
		$\delta^{13}C$	δD	$\delta^{37}Cl$
BTEX				
Benzene	Oxic	−1.5 to −3.5	−11 to −12	
	Anoxic	−1.9 to −3.6	−60 to −79	
Ethylbenzene	Anoxic	−2.1 to −2.2		
Toluene	Oxic	−0.4 to −3.3		
	Anoxic	−0.5 to −2.2	−198 to −728	
m-Xylene	Oxic	−1.7		
	Anoxic	−1.8		
p-Xylene	Oxic	−2.3		
o-Xylene	Anoxic	−1.1 to −3.2		
PAHs				
Naphthalene	Oxic	−0.1		
	Anoxic	−1.1		
Chlorinated hydrocarbons				
Tetrachloroethene	Anoxic	−5.5		−10
Trichloroethene	Oxic	−1.1 to −18.2		
	Anoxic	−2.5 to 13.8		−5.5 to −30
Dichloroethenes	Anoxic	−3.5 to −30.3		
Vinylchloride	Oxic	−1.1 to −11.4		
	Anoxic	−31 to −45		
Dichloromethane	Oxic			−3.8
1,1,2-Trichloroethane	Anoxic	−2.0		
1,2,4-Trichlorobenzene	Anoxic	−3.2 to −3.5		
Fuel oxygenates				
MTBE	Oxic	−1.0 to −2.4	−29 to −66	
	Anoxic	−4.2 to 14.2		
TBA	Oxic	−16.8		

Data adapted and summarized from Meckenstock et al. (2004) and references therein

dilution or degradation. Here, the carbon isotope composition was used to clarify the processes. A principal shift toward lighter isotope ratios is evident with the increase in concentration. This directly points to a main contribution of biotic degradation resulting in decreased pollution, meaning an in situ bioremediation is obvious. Interestingly, the measurements have been individually performed for α-, β- and γ-isomers. Here, different extend of shifts are visible with distance. Moreover, a higher variation for the pesticide isomer γ-HCH was detected, but this variation is clearly higher compared to the differences detected in several technical products (see Fig. 8.6).

As a final example, a complex study combining different approaches to environmental analysis is reported here. The old burden of DDT residues has affected groundwater systems in a complex way. Industrial waste has been dumped, and

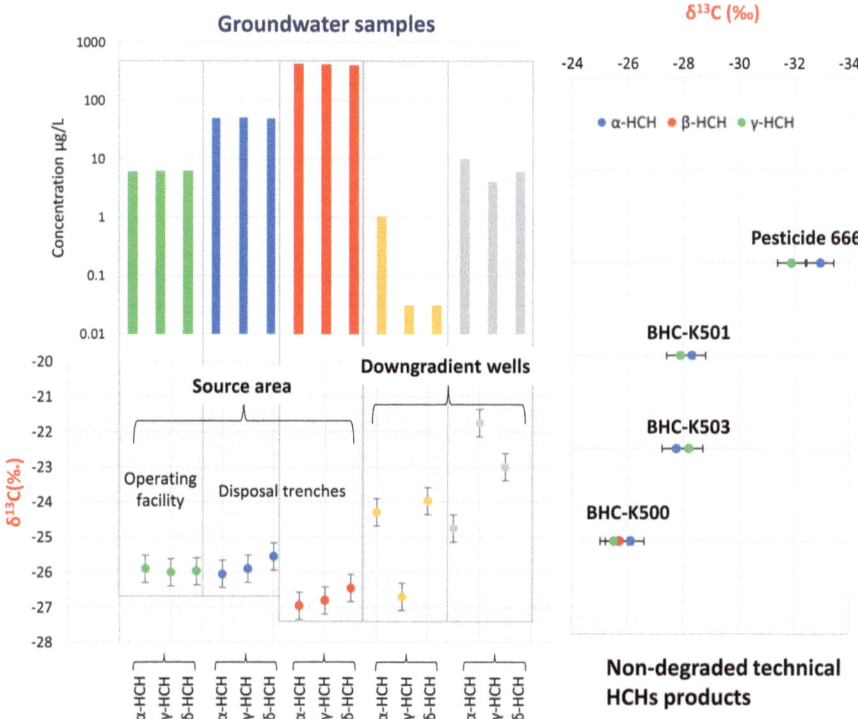

Fig. 8.6 Carbon isotope analyses of HCH isomers in a contaminated aquifer and in technical products. Generally, the decrease of concentrations in the groundwater wells is correlated with the shifts towards heavier isotopic compositions (adapted from and modified after Chartrand et al. 2015)

leakage water has reached the groundwater systems. Besides the covered groundwater directly below the old burden, elevated concentrations of the water-soluble metabolite DDA were detected in an aquifer used for drinking water and directly located at a nearby canal (an infiltration area). Here, the question arose as to whether this second area of contamination resulted from an insufficient barrier in the primary pollution area. This information is essential for any effective protection measures. A first isotopic fingerprinting showed that the carbon isotope ratios varied in both groundwater systems by approx. 3‰, meaning a direct emission pathway was implausible.

The final elucidation of the pathway included two further general aspects of organic environmental pollution, the biotic degradation and the formation of non-extractable residues NER (see also chapter 1.5 in Schwarzbauer and Jovančićević 2018). The DDT contamination was not located at the production site, but effluents were also evident in the nearby canal system. Here, the more lipophilic DDT pollution was associated with the sediments. The degradation pathway of DDT in such riverine sediments under anoxic conditions is well known and shown in Fig. 8.7.

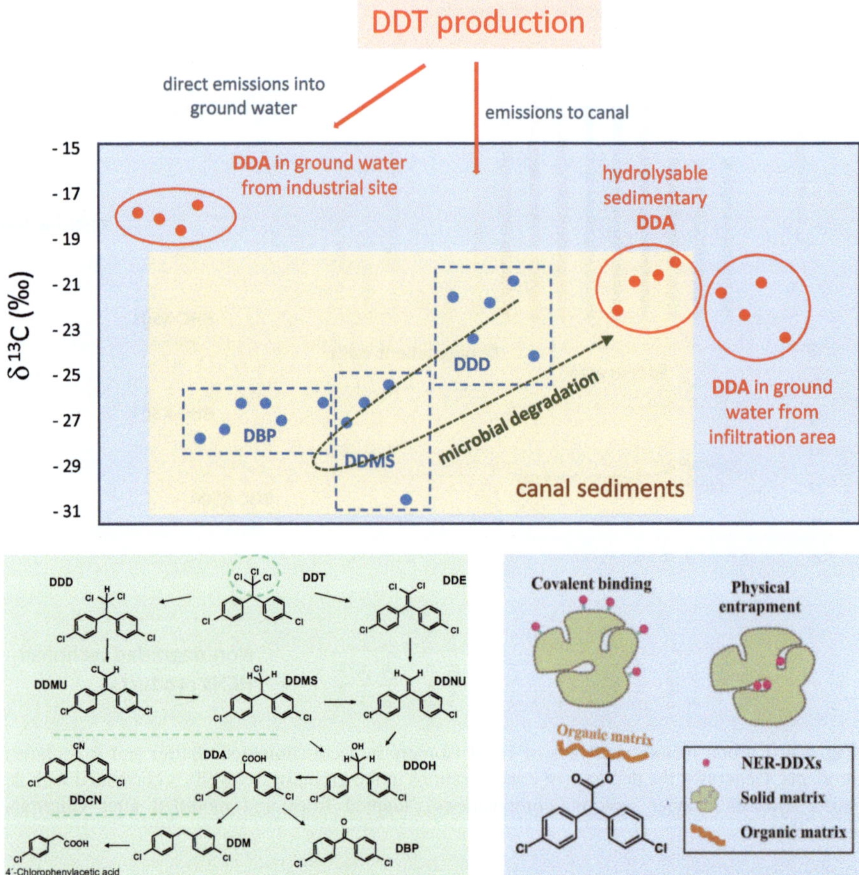

Fig. 8.7 The environmental pathway of a DDX contamination resulting in contamination of an infiltration groundwater area via polluted river sediments. Biodegradation (anaerobic biodegradation pathway in the lower left part) and the formation of non-extractable residues (schematically given in the lower right part) are involved and can be followed by the stable carbon isotope characteristics of all relevant DDT metabolites (according to Schwarzbauer, unpublished results, some parts adapted from Zhu et al. 2019)

DDA, the first water-soluble metabolite, appears after several degradation steps. It is also known that organic pollutants form within soil and sediment organic matter non-extractable residues (bound residues) by adsorption, inclusion or even covalent linkages. Following the DDX (DDT and its metabolite) pollution in the canal along the degradation pathway and including also the bound fraction, the isotopic characterization clearly points to the indirect pathway through the sediments with degradation and formation of NER with subsequent remobilization. This is evident due to the similarity of the carbon isotopic composition of bound sedimentary DDA and the dissolved DDA in the infiltration area.

General Note
Compound-specific isotope analysis is a powerful tool in Environmental Organic Geochemistry, not only as a complementary fingerprinting tool but especially for following biotic transformations in complex ecosystems.

References

Chartrand M, Passeport E, Rose C, Lacrampe-Couloume G, Bidleman TF, Jantunen LM, Lollar S (2015) Compound specific isotope analysis of hexachlorocyclohexane isomers: a method for source fingerprinting and field investigation of *in-situ* bioremediation. Rapid Commun Mass Spectrom 29:505–514

Filippini M, Nijenhuis I, Kümmel S, Chiarini V, Crosta G, Richnow HH, Gargini A (2018) Multi-element compound specific stable isotope analysis of chlorinated aliphatic contaminants derived from chlorinated pitches. Sci Total Environ 640–641:153–162

Griebler C, Safinowski M, Vieth A, Richnow HH, Meckenstock RU (2004) Combined application of stable carbon isotope analysis and specific metabolites determination for assessing in situ degradation of aromatic hydrocarbons in a tar oil-contaminated aquifer. Environl Sci Technol, 38:617–631

Meckenstock RU, Morasch B, Griebler C, Richnow HH (2004) Stable isotope fractionation analysis as a tool to monitor biodegradation in contaminated acquifers. J Contam Hydrol 75:215–255

Morasch B, Richnow HH, Schink B, Meckenstock RU (2001) Stable hydrogen and carbon isotope fractionation during microbial toluene degradation: mechanistic and environmental aspects. Appl Environl Microbiol 67:4842–4849

O'Malley VP, Abrajano TA, Hellou J (1994) Determination of the 13C/12C ratios of individual PAH from environmental samples: can PAH sources be apportioned? Organ Geochem 21:809–822

Schwarzbauer J, Jovančićević B (eds) (2018) Fundamentals in organic geochemistry, vol 3: Organic pollutants in the geosphere. Springer, Cham. 186 pp, ISBN 978-3-319-68937-1

Slater GF, Sherwood Lollar B, King RA, O'Hannesin S (2002) Isotopic fractionation during reductive dechlorination of trichloroethene by zero-valent iron: influence of surface treatment. Chemosphere 49:587–596

Sturchio NC, Clausen JL, Heraty LJ, Huang L, Holt BD, Abrajano TA (1998) Chlorine isotope investigation of natural attenuation of trichloroethene in an aerobic aquifer. Environ Sci Technol 32:3037–3042

Yanik PJ, O'Donnell TH, Macko SA, Qian Y, Kennicutt MC (2003) Source apportionment of polychlorinated biphenyls using compound specific isotope analysis. Org Geochem 34:239–251

Zhu X, Dsikowitzky L, Kucher S, Ricking M, Schwarzbauer J (2019) Formation and fate of point source nonextractable DDT-related compounds in their environmental-terrestrial pathway. Environ Sci Technol 53:1305–1314

Further Reading

Badea SL, Danet A-F (2015) Stable isotope analysis (ESIA)—A new concept to evaluate the environmental fate of chiral organic contaminants. Sci Total Environ 514:459–466

Elsner M, Imfeld G (2016) Compound-specific isotope analysis (CSIA) of micropollutants in the environment—current developments and future challenges. Curr Opin Biotechnol 41:60–72

Kuntze K, Eisenmann H, Richnow HH, Fischer A (2020) Compound-specific stable isotope analysis (CSIA) for evaluating degradation of organic pollutants: an overview of field case studies. In: Boll M (ed) Anaerobic utilization of hydrocarbons, oils, and lipids, Handbook of hydrocarbon and lipid microbiology. Springer, Cham. https://doi.org/10.1007/978-3-319-50391-2

Schmidt TC, Jochmann MA (2012) Origin and fat of organic compounds in water: Characterization by compound-specific stable isotope analysis. Annu Rev Anal Chem 5:133–155